→ INTRODUCING

PARTICLE PHYSICS

TOM WHYNTIE & OLIVER PUGH

Published in
the UK and the USA
in 2013 by Icon Books Ltd,
Omnibus Business Centre,
39–41 North Road, London N7 9DP
email: info@iconbooks.net
www.introducingbooks.com

Sold in the UK, Europe and Asia
by Faber & Faber Ltd,
Bloomsbury House,
74–77 Great Russell Street,
London WC1B 3DA or their agents

Distributed in South Africa
by Book Promotions,
Office B4, The District,
41 Sir Lowry Road,
Woodstock 7925

Distributed in Australia
and New Zealand
by Allen & Unwin Pty Ltd,
PO Box 8500,
83 Alexander Street,
Crows Nest, NSW 2065

Distributed to the trade in the USA
by Consortium Book Sales
and Distribution
The Keg House,
34 Thirteenth Avenue NE, Suite 101,
Minneapolis, MN 55413-1007

Distributed in Canada
by Penguin Books Canada,
90 Eglinton Avenue East,
Suite 700, Toronto,
Ontario M4P 2Y3

ISBN: 978-184831-589-1

Edited by Duncan Heath

Printed and bound in the UK by Clays Ltd, St Ives plc

What are we made of?

It's probably fair to say that we may live a reasonably enjoyable, profitable, and/or meaningful life without knowing the answer to the question: "What is a quark?"

YOU DON'T NEED TO KNOW THE DIFFERENCE BETWEEN A MUON AND A GLUON TO ORDER AND ENJOY A PINT OF BEER, A GLASS OF WINE OR AN ICE-COLD SOFT DRINK.

AN UNDERSTANDING OF ČERENKOV RADIATION WON'T HELP YOU NAVIGATE THE LONDON UNDERGROUND SYSTEM ... AND (FOR BETTER OR WORSE) A FIRM GRASP OF QUANTUM ELECTRODYNAMICS ISN'T REQUIRED TO PROCREATE.

However, if you should stop to think …

… you will have inadvertently embarked on one of the greatest intellectual, philosophical and scientific journeys it is possible to make.

Do not be alarmed. You are not the first to do this, and you will not be the last. It's an age-old endeavour that has baffled, tantalized and inspired some of the finest minds of humanity. It's a quest that has taken us from humble wooden benches in mercury-drenched laboratory sheds to death-defying balloon rides. It has driven us up mountains and down mines. Ultimately, it has led us to construct the enormous laboratory-temples of the Large Hadron Collider.

IT'S THE SEARCH FOR THE ANSWER TO THE QUESTION: "WHAT ARE WE MADE OF?" THIS IS *PARTICLE PHYSICS*.

Philosophy: mind and matter

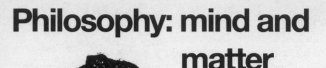

Traditionally, questions like "What are we made of?" were the domain of philosophers. A famous early attempt at an answer can be found in Plato's *Timaeus* (c. 360 BC).

EVERYTHING IS MADE OF FOUR ELEMENTS: *EARTH, AIR, FIRE* AND *WATER.*

Furthermore, these elements were thought themselves to be made of the Platonic solids (the friendliest of the shapes). In Plato's theory of everything, earth is made of stackable cubes, the relative compactness of the octahedron lends itself naturally to the air we find all around us, icosahedra flow much as we would expect water to, and the sharpness of the tetrahedron neatly explains why fire hurts when we touch it. (A fifth element, the "aether", was added by Aristotle to give perfect, unspoilable substance to the heavens.)

Such a theory may well seem like it was formulated in a pub (or the classical equivalent), but even as late as the 18th century, ideas such as Descartes' **Dualism** (*La Description du Corps Humain*, 1647) and Leibniz's **Monads** (*La Monadologie*, 1714) persisted as seemingly reasonable attempts to describe reality.

DUALISM IS THE BELIEF THAT THE MIND AND BODY ARE SEPARATE ENTITIES.

WHEREAS, IN *MONADOLOGIE*, I DESCRIBE THE UNIVERSE AS A NUMBER OF FUNDAMENTAL, IRREDUCIBLE AND INDEPENDENT ELEMENTS THAT EACH THEMSELVES MIRROR THE WHOLE. MY MONADS WERE SOMETHING TO BEHOLD.

Metaphysics

With his elements, Plato was trying to understand what the world was made of. Descartes' Dualism went further, arguing that the stuff that lets us think is different to the stuff we're made of. This division of all things into mind or matter is a great example of **metaphysics** – the branch of philosophy that aims to describe and understand all the aspects of what it means to "be".

THE SUPPOSED UNIFICATION OF MIND AND MATTER THROUGH THE EXISTENCE OF ETERNAL, IRREDUCIBLE, NON-INTERACTING MICROCOSMS REFLECTING THE ENTIRE MACROCOSM IS ALSO AN EXAMPLE OF METAPHYSICS.

And as long as all you're doing is a little postulation and pontification, there's nothing wrong with that.

Empiricism

It was with the birth of John Locke's **empiricism*** in the 17th century that thinkers started acknowledging that checking one's ideas against *experience* might be worthwhile.

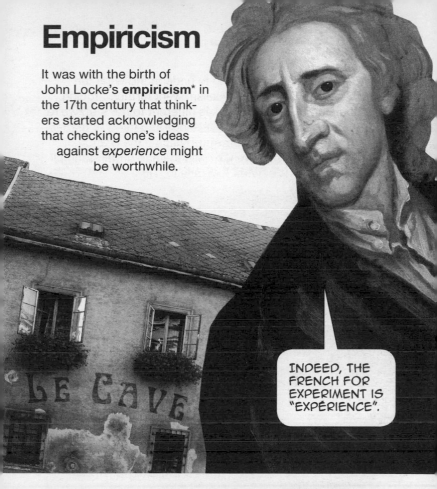

INDEED, THE FRENCH FOR EXPERIMENT IS "EXPÉRIENCE".

This conforms to our modern definition of science and the practice of the scientific method. However, until the 19th century "science" simply meant "knowledge". The term "natural philosophy" was used to describe purely theoretical musings on the workings of the world.

* Words marked with an asterisk are explained in the Glossary on pages 189–90.

Experimental philosophy

Lord Kelvin (1824–1907, born William Thomson) established the first university physics laboratory in Scotland, the spiritual home of empiricism. Here, ideas could be tested scientifically.

IN 1867, I WROTE MY *TREATISE ON NATURAL PHILOSOPHY* WITH PETER GUTHRIE TATE, WHICH SET THE STAGE FOR MUCH OF MODERN PHYSICS.

We have come a long way since the time of brilliant individuals working in what were little more than sheds in the grounds of universities. Experiments at the frontiers of our knowledge now need investments of millions – if not billions – of dollars in Bond villain-esque facilities and equipment, and world-wide networks of computing power for data storage and processing.

And yet, in many ways, modern particle physics retains the spirit of metaphysics. It probes our concept of what is real. One may complain that it's unfair to stunt the creativity of the human imagination by testing its musings against something as trivial as "reality". I prefer to think of it just as working out, as best we can, what's really going on. So far, by testing our ideas with experiments, we have witnessed the triumph of matter over mind.

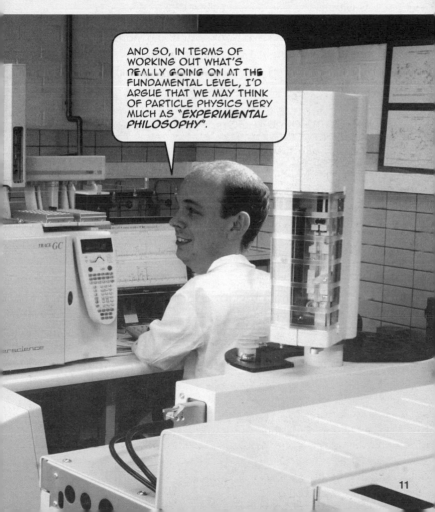

AND SO, IN TERMS OF WORKING OUT WHAT'S REALLY GOING ON AT THE FUNDAMENTAL LEVEL, I'D ARGUE THAT WE MAY THINK OF PARTICLE PHYSICS VERY MUCH AS *EXPERIMENTAL PHILOSOPHY*.

Figuring out the code

The great **Richard Feynman** (1918–88), who shared a Nobel Prize in Physics for his contributions to experimental philosophy, once described science as like trying to figure out the rules of chess by watching a game being played.

The journey described in this book is perhaps more akin to that of a group of characters in a computer game.

THE OBJECTIVE ISN'T TO WIN, OR TO DEFEAT SOME GREAT NEMESIS, OR TO GET THE HIGHEST SCORE.

THE CHARACTERS ARE TRYING TO FIGURE OUT THE COMPUTER CODE USED TO DETERMINE THEIR OWN BEHAVIOUR AND THAT OF EVERYTHING AROUND THEM.

But in terms of this book's subject matter, I'd expand on the analogy a little. The characters can go further. They can ask:

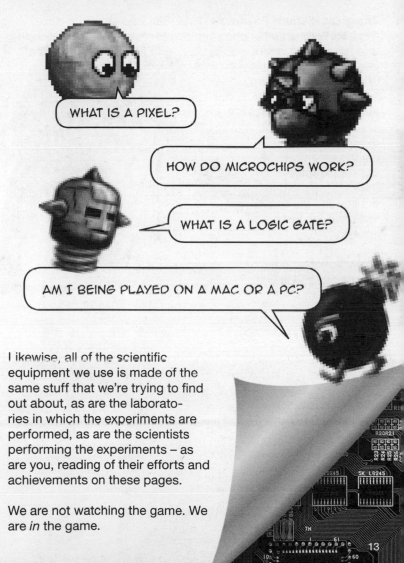

WHAT IS A PIXEL?

HOW DO MICROCHIPS WORK?

WHAT IS A LOGIC GATE?

AM I BEING PLAYED ON A MAC OR A PC?

Likewise, all of the scientific equipment we use is made of the same stuff that we're trying to find out about, as are the laboratories in which the experiments are performed, as are the scientists performing the experiments – as are you, reading of their efforts and achievements on these pages.

We are not watching the game. We are *in* the game.

13

The first atomist

Where did our journey begin? I'd argue that "particle physics" began when we figured out that the **atom** – the indivisible unit of stuff of which all things are made – is, in fact, divisible. To appreciate the seismic shift this represented in our thinking, we need to understand the theory itself and the historical context.

We've already come across one of the first theories of matter – the elements of Plato's *Timaeus*.

UNSURPRISINGLY, THIS DIDN'T STAND UP TO SCIENTIFIC SCRUTINY.

You are probably more familiar with the atomic theory attributed to Thracian philosopher **Democritus** (460–370 BC).

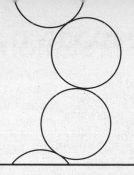

THE UNIVERSE IS COMPOSED OF A MULTITUDE OF TINY, UNCUTTABLE "ATOMS" (*ATOMOS* MEANING UNCUTTABLE IN GREEK), THE PROPERTIES AND INTERACTIONS OF WHICH EXPLAIN THE BEHAVIOUR OF EVERYTHING WE SEE AROUND US.

Democritus was the first **atomist** and, as we shall see, his idea has successfully weathered the tests of both time and experiment. The alternative viewpoint holds that matter is a continuum: one may keep dividing stuff into smaller and smaller pieces *ad infinitum*. It was only towards the end of the 18th century that people started to perform experiments that would shed light onto the matter, as we'll see next.

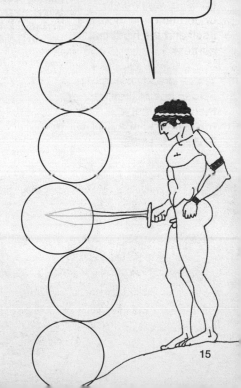

15

The modern atomic theory

Building on the work of French chemists Antoine Lavoisier and Joseph Proust, **John Dalton** (1766–1844) was arguably the father of the modern atomic theory. In *A New System of Chemical Philosophy* (1808), he summarized the results of his experimental work, proposing that each chemical element is made up of atoms with unique properties, and that these atoms combine to form chemical compounds.

THIS ENABLED ME TO MAKE ESTIMATES OF THE ATOMIC WEIGHTS OF THE DIFFERENT ELEMENTS, WITH HYDROGEN TAKING THE DE FACTO LIGHTEST WEIGHT OF "1".

He didn't get it quite right, though. **Joseph Louis Gay-Lussac** (1778–1850) showed that gases combine in whole-number ratios, and that the volumes of gaseous products of such reactions obey similar relationships.

Amedeo Avogadro (1776–1856) seized on this observation to make a big leap forward in how we understand gases. Gay-Lussac's result – that *two* volumes of gaseous water (steam), and not *one*, are formed when hydrogen and oxygen are combined – could be explained only if a given volume of gas contains a fixed number of particles (Avogadro's Law). And gases like hydrogen and oxygen are actually **diatomic** in nature – they are formed of **molecules**[*] each containing two atoms.

Avogadro

Problems with the theory

Despite these successes, the scientific community of the time had a real problem with the atomic hypothesis: they just didn't like what they couldn't see. **Sir Humphry Davy** (1778–1829) – he of the Davy coal-miners' lamp – upon presenting Dalton with a prize at the Royal Society in 1826, remarked that Dalton's atomism didn't really matter as the practical results of his work were still useful.

Unfortunately, improvements in experimental accuracy actually hampered the acceptance of atomic theory.

AS THE MEASUREMENTS GOT BETTER, IT WAS REALIZED THAT THE RATIOS OF THE DIFFERENT ELEMENTS FOUND IN COMPOUNDS WERE NOT QUITE WHOLE NUMBERS AFTER ALL.

ATOMIC WEIGHT OF OXYGEN, CHEMICAL METHOD, 15.879

ATOMIC WEIGHT OF OXYGEN, PHYSICAL METHODS, 15.879

This flew in the face of the notion that the elements were combining in discrete units. We will iron out this conceptual crease later.

It would not be until 1908 – 100 years after the publication of Dalton's seminal work – that experimental evidence supporting **Albert Einstein**'s 1905 explanation of botanist **Robert Brown**'s observation of 1827 would finally convince most people of the existence of atoms and molecules.

Brown (1773–1858) noticed that tiny specks of plant matter spat out by pollen grains suspended in water would randomly jostle about under the lens of his microscope. This became known as Brownian Motion. Einstein (1879–1955) explained this by calculating the distance that the ejected specks would travel if they were being constantly bombarded by the fluid's constituent molecules.

James Clerk Maxwell (1831–1879) and later **Ludwig Boltzmann** (1844–1906) had likewise had much success in describing the thermodynamic behaviour of gases by modelling them as collections of tiny, fast-moving billiard balls continually bashing the surfaces of whatever vessel they were in. This was the **kinetic** theory of gases, and it could be used to explain things like pressure and heat. These billiard balls were, of course, Avogadro's molecules.

But it is one of Maxwell's other great achievements – the unification of electricity and magnetism – that allowed scientists to better understand another aspect of the world around us: **light**.

Some light entertainment

Optics, the science of light and vision, had an early champion in the Islamic scholar **Ibn al-Haytham** (965-1040). His *Book of Optics* (1011–21) succeeded Ptolemy's work on the subject, featuring many experimental results concerning the behaviour of light. It was translated into Latin and influenced later European thinkers including Bacon, Kepler and Descartes.

But while the optics of al-Haytham (or Alhacen, if you prefer the Latin) described what light did, attempts to work out what it actually *was* led to the first appearance of the physicist's favourite question:

IS IT A *PARTICLE* OR IS IT A *WAVE?*

Descartes thought that light was a wave, making an analogy with sound. **Robert Hooke** (1635–1703) also discussed a wave theory of light in *Micrographia* (1665), as did **Christiaan Huygens** (1629–95) in *Treatise on Light* (1690).

However, **Isaac Newton** (1642–1727) – standing on the shoulders of **Pierre Gassendi** (1592–1655) who was, incidentally, an atomist – declared in his *Hypothesis of Light* (1675) that light was made of tiny particles moving in straight lines.

I CALL THESE PARTICLES "CORPUSCLES".

Thanks to Newton's reputation, the corpuscular theory of light dominated 18th-century thinking on the subject.

The conundrum of waves

Ultimately, not even Newton's reputation could withstand the power of the scientific method: the experiments of Thomas Young, Augustine Fresnel, Siméon Poisson and others showed that light reflected, refracted and interfered with itself in a way that could only be explained if light was a wave.

> FURTHERMORE, IT WAS SHOWN THAT LIGHT MUST BE A *TRANSVERSE* WAVE – THE VIBRATIONS WERE SIDE-TO-SIDE, NOT FORWARD-TO-BACK LIKE THE *COMPRESSION* WAVES OF SOUND.

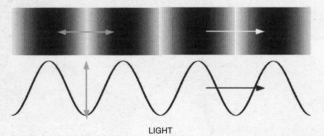

SOUND

LIGHT

But there was a conundrum: waves represent the movement of energy within a medium with no net movement of the medium in question. In other words, when you have waves, you need something to be doing the waving. Light was known to be able to travel in a vacuum – so, unlike with sound, it wasn't air that was carrying the light vibrations. So what was?

The luminiferous aether

Huygens saw this supposed problem with the wave nature of light. Unhappy with the concept of instantaneous "action at a distance", he was convinced that there must be something that was carrying the vibrations of the light waves.

BUILDING ON DESCARTES' CONCEPT OF THE *PLENUM*, I IMAGINED A FORM OF MATTER THAT PERMEATED ALL SPACE: THE *AETHER*.

Newton, of course, wasn't a fan, even though he did employ something like an aether – an "Aethereal Medium" – to explain the refraction of light, something his corpuscles were not equipped to do.

As the wave theory of light took hold, so the "luminiferous aether" seeped into the consciousness of scientists. As more people thought about this aether, it was realized that it must be a very odd substance indeed.

To the vibrations of light, it had to behave as a solid. Transverse vibrations require a solid to move through, and the high frequencies of visible light would require that solid to be diamond-hard. And yet, to slow-moving planets, people, kittens, etc. it behaved as a liquid – as otherwise we'd notice it when moving around.

SO IT WAS LIKE AN ALL-PERMEATING, INVISIBLE SUPER-CUSTARD. (CUSTARD IS A LIQUID IF YOU'RE MOVING SLOWLY, BUT HIT IT HARD ENOUGH AND IT BEHAVES LIKE A SOLID.)

As ridiculous as this sounds, respectable scientists continued to believe in it. Lectures were given, textbooks were written – notably Joseph Larmor's *Aether and Matter* (1900) – and one Victorian "Theory of Everything" even postulated that atoms themselves were nothing but vortices in the aether. One could dedicate an entire book to the topic, but for an excellent summary of how the super-custard was disposed of, I recommend *Introducing Relativity* (spoiler: it's to do with Einstein's Special Relativity*).

THE AETHER SEEMS ODD TO US NOW, A CURIOUS RELIC OF OLD-SCHOOL MECHANICS AND OPTICS. BUT THE MECHANICAL VIBRATION MODEL WAS THE PARADIGM OF THE TIME – PEOPLE SIMPLY COULDN'T IMAGINE LIGHT BEHAVING IN ANY OTHER WAY.

Cathode rays

Certain advances in technology are intrinsically linked with certain advances in science. Astronomy had the telescope. Particle physics had the vacuum tube. Invented by **J.H.W. Geissler** (1814–79) in 1857, along with its successors it enabled scientists to reliably remove all of the matter from a volume of space (i.e. the contents of the tube).

AFTER ALL, IF YOU WANT TO STUDY MATTER AND ITS PROPERTIES, IT'S IMPORTANT TO REMOVE ANY EXTRANEOUS MATTER THAT MIGHT BE GETTING IN THE WAY.

It was the use of vacuum tubes that led to the discovery of the phenomenon that would on the one hand momentarily muddy the waters surrounding the aether debate, but on the other ultimately take science inside the atom: **cathode rays**.

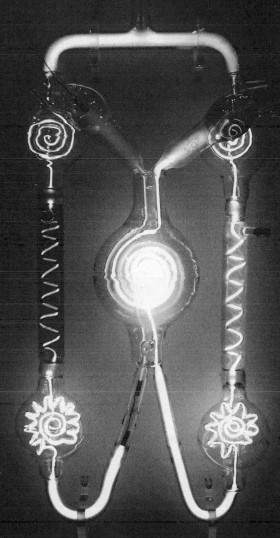

Before that point, **Michael Faraday** (1791–1867) had passed an electric current through a glass jar in order to study electricity in the absence of air. Armed with only a cork stopper and what was basically nothing more than a bicycle pump, he was able to remove enough air from the jar to produce what looked like a beam of light passing between the negatively-charged **cathode** and the positively-charged **anode**.

Geissler's tubes, at a pressure of about a thousandth of an atmosphere, were found to fill instead with a pleasing glow (the basis of today's neon lights).

A new form of light?

Johann Hittorf (1824–1914), working with Geissler's vacuum tubes, noticed in 1869 that the glow seemed to emerge from the cathode in a straight line. **Eugen Goldstein** (1850–1930) christened Hittorf's discovery "cathode rays" in 1876 (*Kathodenstrahlen*). The German school (which also included Heinrich Hertz) held that these strange, glowing rays must be a perturbation in the luminiferous aether.

THESE RAYS MUST BE A NEW FORM OF LIGHT.

Hertz

Meanwhile, across the English Channel, **William Crookes** (1832–1919) had developed a vacuum tube that achieved the pressure required (about one millionth of an atmosphere) to remove enough gas to completely darken the contents of the tube itself, and instead cause the glass at the anode end of the tube to glow.

Crookes used his superior vacuum tubes to show that cathode rays must actually be *corpuscular* in nature.

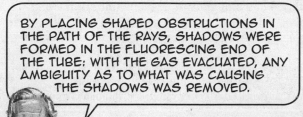

BY PLACING SHAPED OBSTRUCTIONS IN THE PATH OF THE RAYS, SHADOWS WERE FORMED IN THE FLUORESCING END OF THE TUBE; WITH THE GAS EVACUATED, ANY AMBIGUITY AS TO WHAT WAS CAUSING THE SHADOWS WAS REMOVED.

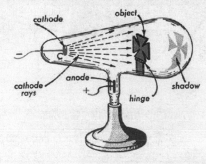

He even placed tiny paddles that could be turned by his hypothe-sized beam of particles when the electric current was switched on.

So who was right? Were cath-ode rays vibrations in the aether, or tiny corpuscles of matter? It would take one of the giants of theoretical and experimental physics to provide the solution.

The Electron

It took the genius of **J.J. Thomson** (1856–1940), one of the last great classical* physicists, to resolve the cathode ray mystery. Through a series of meticulous experiments performed at the University of Cambridge's Cavendish Laboratory, he was able to show that not only were cathode rays particles, but also that these particles were much smaller than the atom. He had discovered the **electron**.

Thomson won the 1906 Nobel Prize in Physics. His Nobel Lecture neatly sums up how it could be shown that the rays observed in Crookes' tubes were negatively-charged particles.

Thomson's proof

Heinrich Hertz (1857–94) had carried out work that led people to believe that this was not the case. This perhaps wasn't surprising, as he believed that the rays were uncharged waves. Thomson refuted Hertz's work.

HERTZ'S VACUUM TUBES WEREN'T GOOD ENOUGH – TOO MUCH GAS WAS PRESENT. THIS GAS WOULD BECOME CHARGED ITSELF, INTERFERING WITH THE EXTERNAL ELECTRIC FIELD. HE WAS SEEING WHAT HE WANTED TO SEE.

MY EQUIPMENT WAS SUPERIOR AND I OBTAINED THE CORRECT RESULT – THE RAYS WERE INDEED DEFLECTED BY ELECTRIC FIELDS.

HERTZ ALSO SHOWED THAT THE RAYS COULD PASS THROUGH THIN METAL FOILS – SOMETHING WHICH FURTHER CONVINCED HIM OF HIS WAVE INTERPRETATION. ATOM-SIZED PARTICLES SURELY COULDN'T PASS THROUGH METAL! BUT WHERE HE SAW WAVES AS THE ONLY EXPLANATION, I WAS INSPIRED TO STUDY THE CORPUSCULAR INTERPRETATION FURTHER.

I REALIZED THAT IF THEY WERE CHARGED PARTICLES MOVING WITH A CERTAIN VELOCITY, THE FORCE THEY WOULD EXPERIENCE IN A MAGNETIC FIELD WOULD BE PROPORTIONAL TO THAT VELOCITY, THE ELECTRIC CHARGE OF THE PARTICLE, AND THE STRENGTH OF THE MAGNETIC FIELD. THE FORCE PRODUCED WOULD CAUSE THE RAY TO BE DEFLECTED.

LIKEWISE, IF AN EXTERNAL ELECTRIC FIELD IS APPLIED TO THE STREAM OF PARTICLES, A FORCE WOULD BE EXPERIENCED BY THE PARTICLES THAT IS PROPORTIONAL TO THE ELECTRIC CHARGE OF THE PARTICLE AND THE STRENGTH OF THE ELECTRIC FIELD.

BY APPLYING A MAGNETIC FIELD AND AN ELECTRIC FIELD OF KNOWN STRENGTHS SIMULTANEOUSLY, *SUCH THAT THE FORCES BALANCE AND NO DEFLECTION OF THE RAYS IS OBSERVED*, THE VELOCITY OF THE CHARGED PARTICLES MAY BE INFERRED. I MADE THIS MEASUREMENT AND FOUND IT TO BE MUCH SLOWER THAN THAT OF LIGHT, RULING OUT HERTZ'S INTERPRETATION OF THE RAYS.

BUT I WASN'T FINISHED THERE. BY MATHEMATICALLY TREATING THE RAYS AS PARTICLES, BY MEASURING THE DEFLECTION OF THE RAYS AS A FUNCTION OF THE ELECTRIC FIELD STRENGTH I COULD MEASURE THE RATIO OF THE PARTICLE'S MASS m TO ITS ELECTRIC CHARGE e.

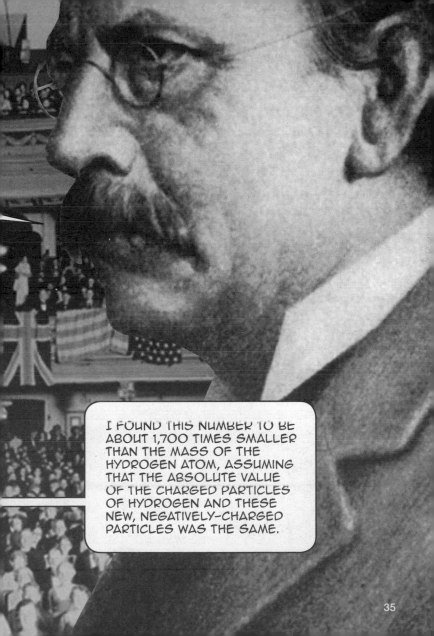

I FOUND THIS NUMBER TO BE ABOUT 1,700 TIMES SMALLER THAN THE MASS OF THE HYDROGEN ATOM, ASSUMING THAT THE ABSOLUTE VALUE OF THE CHARGED PARTICLES OF HYDROGEN AND THESE NEW, NEGATIVELY-CHARGED PARTICLES WAS THE SAME.

THUS IT WAS THAT I SHOWED CATHODE RAYS TO BE COMPOSED OF *CORPUSCLES* WITH A MASS 1,700 TIMES SMALLER THAN THAT OF THE ATOM. AS I REMARKED AT A LECTURE TO THE ROYAL INSTITUTION BACK IN 1897 ... "THE ASSUMPTION OF A STATE OF MATTER MORE FINELY DIVIDED THAN THE ATOM IS A SOMEWHAT STARTLING ONE."

BUT IT WAS THE RIGHT ONE. NOW, WHERE'S MY MEDAL?

"Corpuscles" never really caught on as a name, though, and so another suggestion – "electron" – was adopted for these tiny, negatively-charged particles. The positive particles of hydrogen mentioned above are what we call hydrogen **ions***: atoms or molecules are **ionized*** when they lose or gain electrons.

And so before scientists were even fully convinced of the atomic hypothesis, science had already gone subatomic.

X-rays and radioactivity

Cathode rays were also instrumental in the discovery of another important phenomenon, albeit serendipitously. In 1895, **Wilhelm Röntgen** (1845–1923) was performing an experiment as a follow-up on Hertz's and **Philipp Lenard**'s (1862–1947) work showing that cathode rays could pass straight through thin metal foils.

WORKING IN A DARK-ENED LABORATORY WITH A COVERED VACUUM TUBE, I WAS SURPRISED TO SEE A NEARBY SCREEN PAINTED WITH BARIUM PLATINOCYANIDE LIGHT UP.

While such screens were frequently used as a kind of primitive cathode ray detector, Röntgen was surprised because he wasn't actually using it – it was part of another experiment and well away from the path of the cathode rays under investigation.

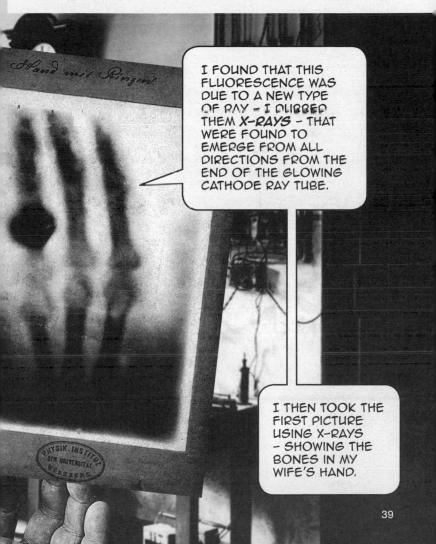

I FOUND THAT THIS FLUORESCENCE WAS DUE TO A NEW TYPE OF RAY – I DUBBED THEM *X-RAYS* – THAT WERE FOUND TO EMERGE FROM ALL DIRECTIONS FROM THE END OF THE GLOWING CATHODE RAY TUBE.

I THEN TOOK THE FIRST PICTURE USING X-RAYS – SHOWING THE BONES IN MY WIFE'S HAND.

Röntgen won 1901's first Nobel Prize in Physics for the discovery of X-rays. These then led (again, a little serendipitously) to the discovery of another important stepping stone: **radioactivity**. Upon hearing of the new X-rays, **Henri Becquerel** (1851–1908) rushed to see if various phosphorescent salts would perform similar feats by exposing them to photographic plates.

MY INITIAL IDEA – THAT THE ENERGY CAME FROM PREVIOUS EXPOSURE OF THE SALTS TO SUNLIGHT – WAS RECTIFIED WHEN I FOUND THAT URANIUM SALTS NEEDED NO SUNLIGHT TO PRODUCE THEIR PENETRATING RAYS.

Becquerel shared the 1903 Nobel Prize in Physics with **Pierre Curie** (1859–1906) and **Marie Curie** (1867–1934) "in recognition of the extraordinary services he has rendered by his discovery of spontaneous radioactivity".

Enter Rutherford

X-rays and radioactivity generated much excitement among scientists and the wider public. People were delighted by the novelties these discoveries afforded – the photographs of their feet bones through their shoes, the glowing hands on watches – blissfully unaware of the darker side of this mysterious light.

It fell to **Ernest Rutherford** (1871–1937) to unravel the nature of these new phenomena. Born and educated in Canterbury, New Zealand, Rutherford joined the Cavendish Laboratory as its first research student (typically one joined having come up through the Cambridge undergraduate system).

AFTER SOME INITIAL WORK ON RADIO WAVES, I DETERMINED (UNDER THE SUPERVISION OF J.J. THOMSON) THAT X-RAYS, TOO, FORM PART OF THE ELECTROMAGNETIC SPECTRUM – *THEY ARE A FORM OF LIGHT.*

Rutherford then showed that Becquerel's rays were in fact made up of three types of **radiation**. Of the three, he identified the alpha and beta types, and named the third type, gamma, which had been discovered by French chemist **Paul Villard** (1860–1934) in 1900. They were distinguished by their different penetrating powers. Generally speaking, "radiation" refers to particles or waves that transmit energy without needing some sort of transmitting medium.

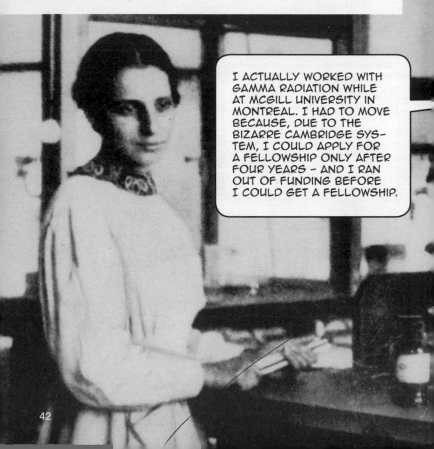

I ACTUALLY WORKED WITH GAMMA RADIATION WHILE AT MCGILL UNIVERSITY IN MONTREAL. I HAD TO MOVE BECAUSE, DUE TO THE BIZARRE CAMBRIDGE SYSTEM, I COULD APPLY FOR A FELLOWSHIP ONLY AFTER FOUR YEARS – AND I RAN OUT OF FUNDING BEFORE I COULD GET A FELLOWSHIP.

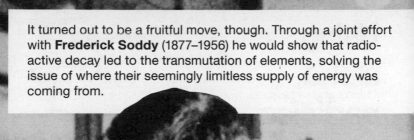

It turned out to be a fruitful move, though. Through a joint effort with **Frederick Soddy** (1877–1956) he would show that radioactive decay led to the transmutation of elements, solving the issue of where their seemingly limitless supply of energy was coming from.

I ALSO DEVELOPED THE CONCEPT OF THE "HALF LIFE" OF A RADIOACTIVE MATERIAL – THE LENGTH OF TIME IT TAKES FOR THE RADIOACTIVITY OF A SUBSTANCE TO HALVE.

He won the 1908 Nobel Prize in Chemistry "for his investigations into the disintegration of the elements, and the chemistry of radioactive substances". But the South Island's most famous son was only just getting started.

Using alpha particles as tools

Rutherford returned to England in 1907 as Langworthy Professor of Physics at the University of Manchester, where he continued his research on radioactivity. He realized that alpha particles, while interesting in their own right, could be used as tools to further probe the structure of matter.

With **Hans Geiger** (1882–1945) he developed the means to exploit his alpha radiation: zinc sulphide screens that would flash (scintillate) and ionization chambers that would allow them to detect and count individual alpha particles.

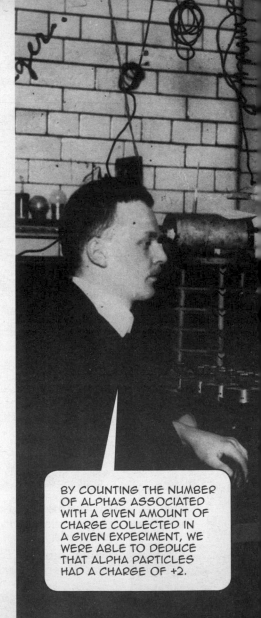

BY COUNTING THE NUMBER OF ALPHAS ASSOCIATED WITH A GIVEN AMOUNT OF CHARGE COLLECTED IN A GIVEN EXPERIMENT, WE WERE ABLE TO DEDUCE THAT ALPHA PARTICLES HAD A CHARGE OF +2.

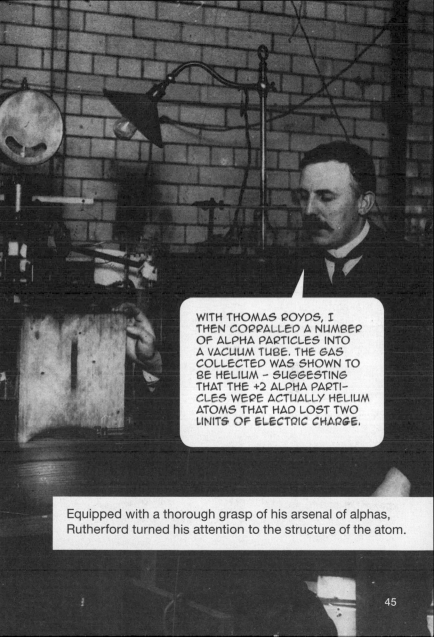

WITH THOMAS ROYDS, I THEN CORRALLED A NUMBER OF ALPHA PARTICLES INTO A VACUUM TUBE. THE GAS COLLECTED WAS SHOWN TO BE HELIUM – SUGGESTING THAT THE +2 ALPHA PARTICLES WERE ACTUALLY HELIUM ATOMS THAT HAD LOST TWO UNITS OF ELECTRIC CHARGE.

Equipped with a thorough grasp of his arsenal of alphas, Rutherford turned his attention to the structure of the atom.

J.J. Thomson had suggested that, in the atom, his negatively-charged electrons (corpuscles) were distributed evenly throughout a sphere of equal positive charge, like plums in a pudding.

I DECIDED TO TEST THIS MODEL BY SHOOTING POSITIVELY-CHARGED ALPHA PARTICLES AT A THIN FOIL OF GOLD.

The even distribution of positive and negative charges in the pudding-like gold atoms would only mildly perturb the path of his positively-charged projectiles. However, initial experiments by Geiger showed that the alphas were being scattered out to wider angles than Thomson's model predicted.

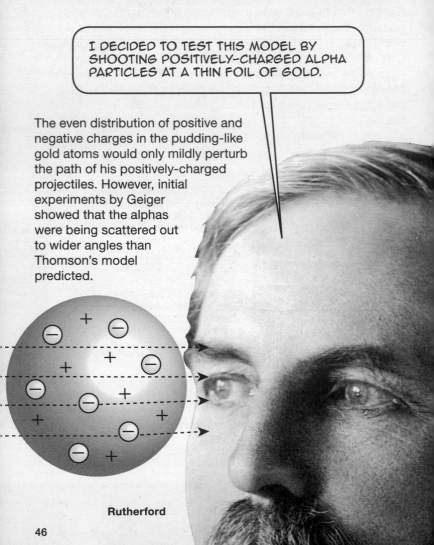

Rutherford

The nuclear model

Sensing that they were on to something, but mindful of wild goose chases and the hefty demands on the schedule of a world-class scientific facility, Rutherford did what all great scientists do even to this day: give the potentially iffy follow-up work to an undergraduate student. The "lucky" student was **Ernest Marsden** (1889–1970) and his results not only confirmed that the alphas were being scattered more widely than expected in the pudding model, but some of the alphas were coming back straight at the source.

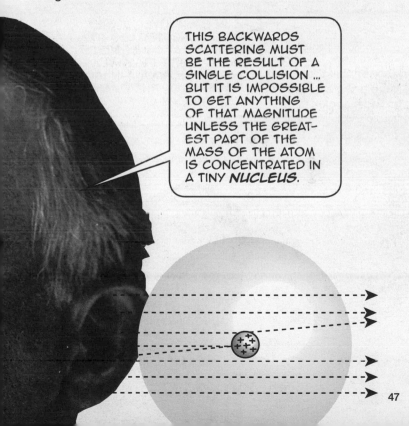

THIS BACKWARDS SCATTERING MUST BE THE RESULT OF A SINGLE COLLISION ... BUT IT IS IMPOSSIBLE TO GET ANYTHING OF THAT MAGNITUDE UNLESS THE GREATEST PART OF THE MASS OF THE ATOM IS CONCENTRATED IN A TINY *NUCLEUS*.

This realization marked the birth of the nuclear model of the atom. Thomson's negatively-charged electrons had been pictured orbiting a tiny, massive, positively-charged nucleus. But there were huge problems with the model: classically speaking, such orbits would result in the energy of the electrons leaking away from the atom almost immediately, leaving them to fall helplessly into the nucleus.

Fortunately for Rutherford, a research student newly arrived in England from Denmark had failed to integrate with J.J. Thomson's Cavendish Laboratory.

A MEETING OVER DINNER IN MANCHESTER LED TO "THE GREAT DANE" *NIELS BOHR* (1885-1962) MOVING TO MY GROUP IN 1912 TO DEVELOP THE QUANTUM THEORY OF THE ATOM.

The rest (as they say) is history: a history that is beautifully told in *Introducing Quantum Theory*.

A light distraction: the photon

Of course, you can't tell the story of particle physics without quantum theory* – and we won't try to. But the story of Bohr, Heisenberg, Schrödinger et al. and their struggle to understand the workings of the atom in terms of the utterly unintuitive quantum theory has been told elsewhere. We will need to weave in a thread from the quantum tapestry, though, as it leads us to another "new" particle: the **photon**.

After performing experiments with cathode rays on metal foils, Philipp Lenard started using beams of light instead, and he found that light could knock electrons out of the foil like the cathode rays (which were themselves electrons) could.

THANKS TO MY CARE-FUL MEASUREMENTS, I NOTICED SOME-THING RATHER ODD ...

Maxwell again

By this point, Maxwell had successfully unified the phenomena of electricity and magnetism in his aptly-titled *A Treatise on Electricity and Magnetism* (1873).

> I DESCRIBED ELECTRICITY AND MAGNETISM USING *FIELDS*. A FIELD IS A CONCEPT IN PHYSICS WHERE EVERY POINT IN SPACE IS ASSIGNED A VALUE. WITH ELECTRIC AND MAGNETIC FIELDS, THIS VALUE CORRESPONDS TO A FIELD STRENGTH WITH A DIRECTION.

Clarendon Press Series

A TREATISE

ON

ELECTRICITY AND MAGNETISM

BY

JAMES CLERK MAXWELL, M.A.

LLD. EDIN., F.R.SS. LONDON AND EDINBURGH
HONORARY FELLOW OF TRINITY COLLEGE,
AND PROFESSOR OF EXPERIMENTAL PHYSICS
IN THE UNIVERSITY OF CAMBRIDGE

VOL. II

Orford

AT THE CLARENDON PRESS

1873

These fields provided the framework and the mathematical toolkit to describe the lines of force that Faraday had previously discussed in his research. Maxwell showed that the speed of the vibrations in these fields was suspiciously close to the measured speed of light – and conjectured that it *wasn't* a coincidence.

Maxwell correctly deduced that light was, in fact, oscillations in these fields that travel at a constant speed, roughly 300,000,000 metres per second. The frequency of the light waves (or, equivalently, the wavelength) corresponded to the colour of the light – red was lower-frequency and a longer wavelength, violet was higher-frequency and so a shorter wavelength. (Rutherford had shown that X-rays were also electromagnetic waves with a frequency much higher than that of visible light.)

Lenard, meanwhile, was experimenting with beams of monochromatic light, using a prism to separate out light with a small range of frequencies.

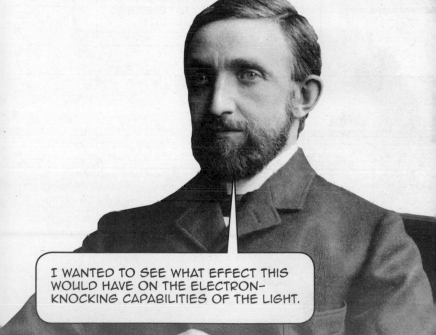

I WANTED TO SEE WHAT EFFECT THIS WOULD HAVE ON THE ELECTRON-KNOCKING CAPABILITIES OF THE LIGHT.

Curiously, Lenard found that using *more* light didn't necessarily mean more electrons were knocked out of the metal – what really mattered was the **frequency** of the light. Below a certain frequency, the electrons wouldn't budge. This "photoelectric effect" couldn't be explained in terms of the classical wave theory.

It was one of Albert Einstein's famous papers from his *annus mirabilis* of 1905 that provided the explanation.

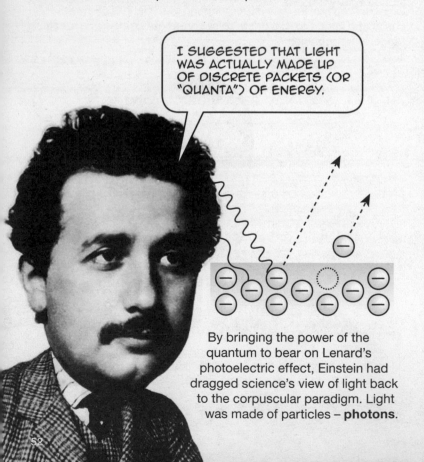

I SUGGESTED THAT LIGHT WAS ACTUALLY MADE UP OF DISCRETE PACKETS (OR "QUANTA") OF ENERGY.

By bringing the power of the quantum to bear on Lenard's photoelectric effect, Einstein had dragged science's view of light back to the corpuscular paradigm. Light was made of particles – **photons**.

Inside the nucleus

With his alpha particles, Rutherford had hit upon (so to speak) the nucleus. But what exactly made up the clump of mass at the centre of the atom? It was known that the properties of the various elements were determined by each element's **atomic number**, Z. Atoms of a given element had Z electrons that could (in principle, if not in practice) be dislodged, leaving an ion* with a charge of +Z.

> HOWEVER, THE MASSES OF EACH ELEMENT – THE *ATOMIC MASS*, A, GIVEN IN UNITS OF THE HYDROGEN ATOM'S MASS – WERE SHOWN TO BE ROUGHLY TWICE Z.

What was more, some elements with the same atomic number Z (and so the same chemical properties) were shown to have differing atomic masses. These were known as **isotopes** (which finally explained the non-whole-number atomic masses that had fuelled earlier objections to the atomic hypothesis). What was going on inside the nucleus?

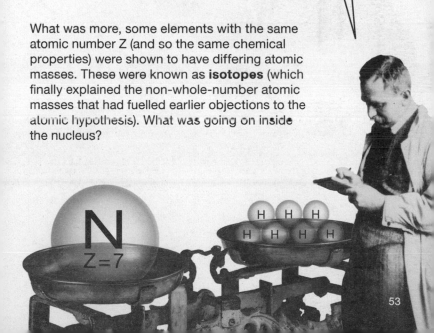

N
Z=7

H H H
H H H H

The proton

Rutherford, on something of a nuclear winning streak, worked out that the nuclei of heavier elements were themselves composed of hydrogen nuclei. You won't win any prizes for guessing that he used alpha particles to do so. Upon finding hydrogen nuclei produced as a result of bombarding air with alphas, he found that hitting nitrogen (a major component of air) with an alpha would produce an ionized hydrogen nucleus and oxygen.

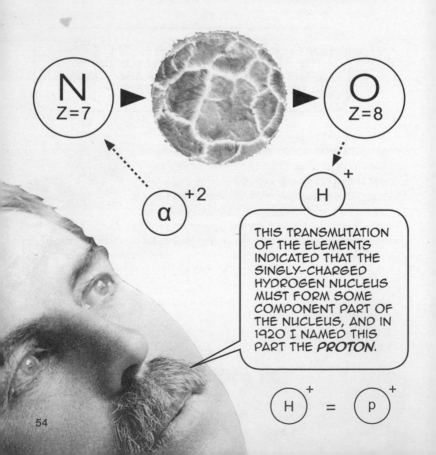

THIS TRANSMUTATION OF THE ELEMENTS INDICATED THAT THE SINGLY-CHARGED HYDROGEN NUCLEUS MUST FORM SOME COMPONENT PART OF THE NUCLEUS, AND IN 1920 I NAMED THIS PART THE *PROTON*.

The nuclei of the different isotopes all had atomic masses in multiples of the proton mass. However, one simply couldn't say that the nucleus contained A protons.

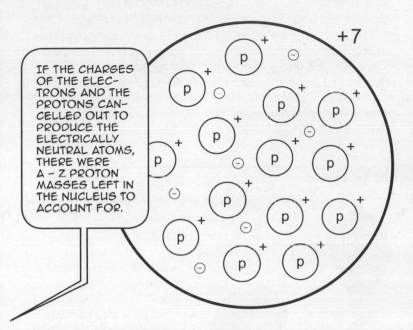

IF THE CHARGES OF THE ELECTRONS AND THE PROTONS CANCELLED OUT TO PRODUCE THE ELECTRICALLY NEUTRAL ATOMS, THERE WERE A – Z PROTON MASSES LEFT IN THE NUCLEUS TO ACCOUNT FOR.

The initial suggestion, that there were an additional A – Z electrons present in the nucleus, was fraught with quantum mechanical problems (confining electrons to such a tiny space would mean they had a huge momentum by Heisenberg's Uncertainty Principle*, and the quantum **spin**[†] of the nitrogen nucleus didn't make sense if it contained 14 protons and 7 electrons). Of course, no one believed in quantum theory anyway, and certainly no one wanted any more new particles to deal with, so it was left at that …

[†] "Spin" is a quantum-mechanical property of a particle that relates to its angular momentum. The name suggests that you can think of the particle spinning about an axis, but the reality is far, far more complicated than that …

The neutron

Then, in 1932, **James Chadwick** (1891–1974) seized upon the results of experiments performed in Germany and France. Walther Bothe and his student Herbert Becker had bombarded beryllium with relatively high-energy alpha particles from polonium and found that a new form of uncharged radiation was released in the ensuing reaction. Meanwhile, Irène and Frédéric Joliot-Curie in France (Irène was the Curies' daughter) had bombarded paraffin with a similar kind of uncharged, high-energy radiation from beryllium itself.

WE INTERPRETED THIS AS A VERY HIGH-ENERGY GAMMA RAY.

But Chadwick, from his previous collaborations with Rutherford, knew what he was looking for and could show that the gamma ray hypothesis was untenable. He performed a series of experiments that confirmed the existence of a new, neutral particle with a mass slightly larger than that of the proton. In 1935 he was awarded the Nobel Prize in Physics for the discovery of the neutron.

The final picture?

The model of the atom was now complete: a tiny nucleus of Z positively-charged protons and A – Z uncharged neutrons (collectively known as the **nucleons***) was orbited by Z electrons. These electrons could jump about between orbits – or leave the atom entirely – by exchanging photons, Einstein's quantized chunks of light.

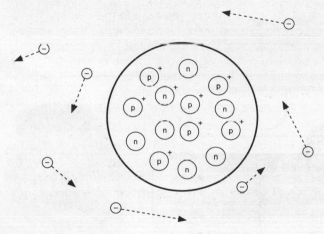

If you closed your mind's eye and believed in the quantum theory of Bohr and his colleagues you could use the model to explain pretty much all of chemistry. And with chemistry, you can build up molecules – molecules like the amino acids required for DNA and all the wonders of life.

So why would you need anything more than protons, neutrons, electrons and photons? Why did particle physics need to go beyond the basic constituents of the atom? As you'll see in the rest of the book, it's because we had to. To find out why, we'll need to back up a bit to the first decade of the 20th century.

The discovery of cosmic rays

With the invention of the vacuum tube, scientists were provided with a way of simplifying the physical system they were studying by removing pretty much everything else from a fixed volume of space. Experimental physics requires a firm control of your environment; what you can't control, you have to make adjustments for when processing your results. Good experimental physicists will do this when analysing their experiment for systematic errors. The problem comes when your source of interference is something far beyond what the scientific paradigms of the time will allow you to imagine.

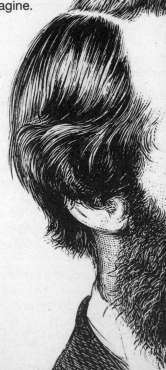

THAT SAID, SUCH ANNOYANCES (OBSERVED BY THE RIGHT EYES) ARE OFTEN GOLDMINES FOR REVOLUTIONARY SCIENTIFIC DISCOVERIES. SUCH WAS THE CASE WITH *COSMIC RAYS*.

Something in the air

As far back as 1785, **Charles-Augustin de Coulomb** (1736–1806) had shown that an isolated, charged body spontaneously loses its charge. Faraday and Crookes went on to show that if you removed the air from around the charged body (using, say, a vacuum tube), the rate at which the charge was lost decreased.

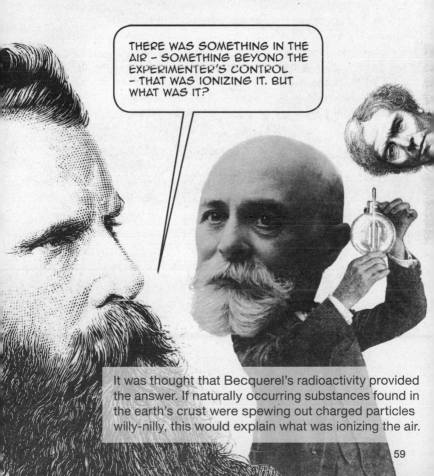

THERE WAS SOMETHING IN THE AIR – SOMETHING BEYOND THE EXPERIMENTER'S CONTROL – THAT WAS IONIZING IT. BUT WHAT WAS IT?

It was thought that Becquerel's radioactivity provided the answer. If naturally occurring substances found in the earth's crust were spewing out charged particles willy-nilly, this would explain what was ionizing the air.

The experimental proof of this common-sense assumption proved tricky to actually find. Better and better electroscopes – devices that could be used to detect electric charge – were being developed at the turn of the century. Among the scientists working in the field at the time, **C.T.R. Wilson** (1869–1959) is of particular note. After developing an interest in chemistry and physics at Cambridge, he found that his real passion was meteorology: the science of weather.

ELECTROSCOPES WERE OF IMMENSE USE IN THE STUDY OF THE ATMOSPHERIC ELECTRICITY ASSOCIATED WITH THUNDERSTORMS.

Wilson's work (among that of others) led to the improved electroscopic measurements that cast doubt on the crusty radiation hypothesis.

An extraterrestrial source?

I WONDERED WHETHER THE SOURCE OF IONIZATION MIGHT NOT BE COMING FROM THE EARTH AT ALL – IF IT MIGHT BE OF SOME EXTRATERRESTRIAL SOURCE, FROM SOMEWHERE ELSE IN THE COSMOS.

But when Wilson tested his hypothesis in the tunnels of the Caledonian Railway in Scotland he found no difference in the rate of ionization.

THE ROCKS CLEARLY WEREN'T ACTING AS A SHIELD, AS I EXPECTED.

THE *ROCKS?* THE POTENTIALLY *RADIOACTIVE ROCKS?*

Wilson gave up on the idea but others didn't. In 1909, **Theodore Wulf** (1868–1946) developed a portable electroscope that enabled him to compare the rate of ionization at the top and bottom of the Eiffel Tower. In 1911, **Domenico Pacini** (1878–1934) went the other way and noted that the ionization rate measured under water decreased (he also found that the rate at sea – where there was no nearby radioactive crust – was the same as that measured on land).

But to be taken seriously in the field of atmospheric electricity, you needed a balloon. This was the nearest people could get to space at the time (rockets were a long way off).

IF YOU WANTED TO MEASURE SOMETHING THAT WAS SUPPOSEDLY COMING FROM OUTER SPACE, YOU'D HAVE TO GET AS CLOSE AS YOU COULD.

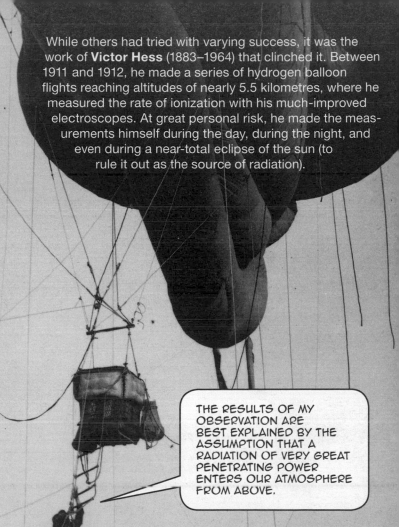

While others had tried with varying success, it was the work of **Victor Hess** (1883–1964) that clinched it. Between 1911 and 1912, he made a series of hydrogen balloon flights reaching altitudes of nearly 5.5 kilometres, where he measured the rate of ionization with his much-improved electroscopes. At great personal risk, he made the measurements himself during the day, during the night, and even during a near-total eclipse of the sun (to rule it out as the source of radiation).

THE RESULTS OF MY OBSERVATION ARE BEST EXPLAINED BY THE ASSUMPTION THAT A RADIATION OF VERY GREAT PENETRATING POWER ENTERS OUR ATMOSPHERE FROM ABOVE.

Hess shared the 1936 Nobel Prize in Physics "for his discovery of cosmic radiation". This was truly the era of the adventuring physicist – the particle-hunter.

The cloud chamber

It's probably fair to say that C.T.R. Wilson wasn't too bothered about missing out on the discovery of cosmic rays. Back in September 1894, when he observed the "wonderful optical phenomena shown when the sun shone on the clouds" on Ben Nevis, he had decided to try to recreate the phenomenon in the laboratory. (As I said, physicists are all about control – and who wouldn't want clouds on demand?) He developed his **cloud chamber** while working in the Cavendish Laboratory.

I COULD MAKE CLOUDS BY *SUPER-SATURATING* MOIST AIR IN A SEALED CHAMBER – PRESSURIZING IT SO THAT ANY WATER VAPOUR PRESENT WOULD CONDENSE IN THE FORM OF DROPLETS AT THE SLIGHTEST DISTURBANCE.

DROPLETS WOULD FORM AROUND ANY DUST IN THE AIR, MAKING THE CLOUDS I WAS HOPING TO CREATE. BUT EVEN AFTER THE DUST HAD ALL GONE, CLOUDS WERE STILL FORMING.

THEN RÖNTGEN DISCOVERED X-RAYS, AND IN 1896 I USED THE CAVENDISH'S OWN X-RAY SOURCE TO ARTIFICIALLY IONIZE THE AIR IN THE CHAMBER AND CREATE CLOUDS AT WILL.

THESE, BY THE WAY, WERE THE SAME DROPLETS I WOULD USE TO MEASURE THE CHARGE ON MY CORPUSCLES, AHEM, ELECTRONS.

J.J. Thomson

Wilson then turned his attention to electroscopes, before returning to "increasing the usefulness of the condensation method" in 1910. To say he succeeded would be something of an understatement. His original plan was to investigate the streams of ions produced via Lenard's photoelectric effect in a cloud chamber.

The American physicist **Robert Millikan** (1868–1953), who in 1908 measured the charge on the electron using drops of oil instead of water, had already successfully investigated Einstein's photon explanation. So Wilson turned to his other goal – using the cloud chamber to visualize the paths of Rutherford's alpha and beta radiation.

I THOUGHT THAT THE CHARGED ALPHAS AND BETAS WOULD CONDENSE THE WATER AS THEY PASSED THROUGH THE CHAMBER – AND WAS DELIGHTED TO SEE *LITTLE WISPS AND THREADS OF CLOUDS* WHEN I TRIED IT.

Wilson was also able to take photographs of the trails left by the radiation. By 1923, he had perfected the first particle tracker, crucial to the study of the newly discovered cosmic rays. He shared the 1927 Nobel Prize in Physics "for his method of making the paths of electrically charged particles visible by condensation of vapour".

Tracking the cosmic rays

Cloud chambers quickly became the "must-have" apparatus for physics laboratories around the world. (Apart from their scientific utility, they also look particularly beautiful – if you ever get the chance, try to see one in operation.) As well as tracking the particles from radioactive sources, they allowed cosmic rays to be studied in detail. **Dmitri Skobeltsyn** (1892–1990) performed many pioneering studies in the field.

AS WE HAVE SEEN, CHARGED PARTICLES MOVING IN A MAGNETIC FIELD EXPERIENCE A FORCE, PROPORTIONAL TO THEIR VELOCITY. THIS CAUSES THEIR TRAJECTORIES TO CURVE.

BY PLACING A CLOUD CHAMBER IN A MAGNETIC FIELD AND PHOTOGRAPHING THE RESULTANT CURVED TRACKS, THE MOMENTUM P OF THE COSMIC RAY PARTICLES CAN BE MEASURED.

Classically, the momentum of a particle is its mass multiplied by its velocity.

THE ORIENTATION OF THE CURVATURE RELATIVE TO THE MAGNETIC FIELD WOULD ALSO TELL YOU THE SIGN OF THE CHARGE – POSITIVE OR NEGATIVE. IN 1929 I NOTICED SOME TRACKS THAT LOOKED LIKE THEY HAD THE MASS OF ELECTRONS BUT WERE CURVING THE WRONG WAY.

At the California Institute of Technology (Caltech), Millikan had established a research group devoted to the study of cosmic rays. Cosmic rays were of great interest because they were a free source of particles with energies far in excess of those available from radioactive sources of the day.

Carl David Anderson (1905–91) was a member of this group who, with Millikan, planned and built a cloud chamber situated in a magnetic field 40,000 times the strength of the earth's. He also placed a 6mm plate of lead between the upper and lower portions of the chamber to reduce the ambiguity surrounding the direction of the photographed tracks: particles would lose energy as they passed through the lead, increasing the curvature as the particle momentum subsequently decreased.

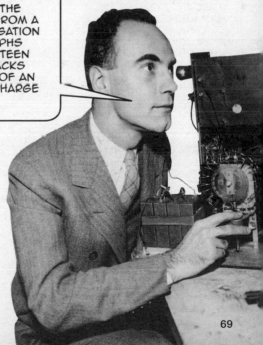

IN 1932 I REPORTED THE RESULTS OBTAINED FROM A PAINSTAKING INVESTIGATION OF 1,300 PHOTOGRAPHS OF COSMIC RAYS. FIFTEEN IMAGES SHOWED TRACKS THAT HAD THE MASS OF AN ELECTRON BUT THE CHARGE OF A PROTON.

This would have flummoxed physicists – to whom the only charged particles in existence were negative electrons and positive protons – were it not for the work of another quantum hero, Paul Dirac.

Quantum field theory

Paul Dirac (1902–84) was an English physicist who had already made waves in 1925 with his reinterpretation of Heisenberg's matrix mechanics* and Schrödinger's wave mechanics* as a special case of a more general mathematical framework. In Copenhagen (and later in Göttingen) he turned his attention to the absorption and emission of electromagnetic radiation.

Born

IN 1925, MAX BORN, PASCUAL JORDAN AND I HAD BROKEN MAXWELL'S CLASSICAL ELECTRO-MAGNETIC FIELDS INTO CHUNKS (OR *QUANTA*) OF A FIXED ENERGY ...

... JUST AS EINSTEIN HAD WITH LIGHT WAVES ...

Jordan

... EXCEPT NOW *EVERY POINT IN SPACE* HAD A CHUNK OF ENERGY ASSOCIATED WITH IT – WHICH WE THOUGHT OF AS AN INFINITE SET OF TINY SPRINGS (HARMONIC OSCILLATORS, TECHNICALLY SPEAKING).

Heisenberg

> I BROUGHT MATTER – SPECIFICALLY, ELECTRONS – INTO THE EQUATION TO PRODUCE THE FIRST COMPLETE *QUANTUM FIELD THEORY* IN 1927.

Dirac

Crucially, in Dirac's theory one could change the number of particles involved in an interaction, something that couldn't be done with, say, Schrödinger's wave equation. Matter could now be created and destroyed with mathematics.

More fundamentally, light and matter now no longer needed to be thought of as either a wave or a particle – *everything* could be thought of as a number of perturbations in an infinite number of sets of an infinite number of tiny springs, oscillating away to represent the movement of energy in the universe in all its wondrous forms.

Dirac's amazing equation

The equal treatment of electrons and photons as quantized fields also sat well with the brilliant suggestion of **Louis de Broglie** (1892–1987).

IN MY PhD THESIS OF 1924 I PUT FORWARD THE HYPOTHESIS THAT MATTER SHOULD ALSO BEHAVE LIKE A WAVE ...

Experimental confirmation of this came from **Joseph Davisson** (1881–1958) and **G.P. Thomson** (1892–1975), son of J.J. Thomson, about 30 years after his father first demonstrated that electrons were particles. But while the wave–particle duality could now be put to one side, a bigger problem was caused by Einstein's Special Relativity.

The equations that the early quantum theorists had been using were not relativistically invariant – you couldn't transform from one frame of reference to another in the way required to keep the speed of light constant (again, see *Introducing Relativity*). The equations weren't the right shape to do so.

ATTEMPTS TO USE EQUATIONS THAT WERE THE RIGHT SHAPE WERE THOUGHT TO BE IMPOSSIBLE – BUT I TRIED ANYWAY.

I FOUND THAT SOLUTIONS WERE POSSIBLE WHEN THE MATRIX ALGEBRA OF QUANTUM MECHANICS WAS EMPLOYED TO DESCRIBE THE QUANTIZED FIELDS.

$$\left(\beta mc^2 + \sum_{k=1}^{3} \alpha_k p_k c\right) \psi(\mathbf{x},t) = i\hbar \frac{\partial \psi(\mathbf{x},t)}{\partial t}$$

Dirac's equation was a staggering achievement for physics. He had united two new and exciting branches of physics: quantum mechanics and Special Relativity. The fact that electrons had quantum-mechanical spin plopped out of the algebra automatically.

Antimatter

Schrödinger's wave equation couldn't achieve what Dirac's could. Dirac's equation also produced the correct value for the electron's **magnetic moment** – the property of the electron that determines the force exerted on it by a magnetic field. But most amazingly, the solution to the Dirac equation suggested that there should be a "mirror" particle to the electron with the opposite charge.

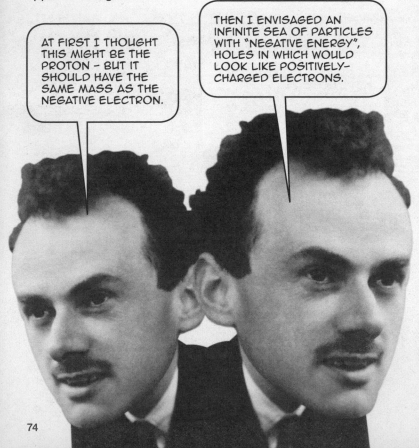

AT FIRST I THOUGHT THIS MIGHT BE THE PROTON – BUT IT SHOULD HAVE THE SAME MASS AS THE NEGATIVE ELECTRON.

THEN I ENVISAGED AN INFINITE SEA OF PARTICLES WITH "NEGATIVE ENERGY", HOLES IN WHICH WOULD LOOK LIKE POSITIVELY-CHARGED ELECTRONS.

Anderson

OF COURSE, I REALIZED THAT THE TRACKS IN MY PHOTOGRAPHS WITH THE ELECTRON'S MASS BUT THE WRONG CURVATURE WERE A NEW PARTICLE IN THEIR OWN RIGHT – ANTI-ELECTRONS, OR *POSITRONS*.

Dirac shared the 1933 Nobel Prize in Physics with Schrödinger "for the discovery of new productive forms of atomic theory", while Anderson took half of the 1936 Prize "for his discovery of the positron" (Hess took the other half). The combination of high-energy cosmic rays, Wilson's cloud chamber and this new picture of light and matter had chalked up the first major departure from "everyday" particles: **antimatter**.

Quantum electrodynamics

Dirac's quantum field theory predicted the existence of antimatter, but it did a lot more than that. The framework gave physicists a new insight into what was going on with the electromagnetic force. For example, when two electrons – both negatively charged – repel each other, this can be thought of in terms of the electrons exchanging momentum by passing photons between them. (These are virtual photons, not real photons; nothing actually lights up.)

The theory also showed that an electron and a positron – matter and antimatter – could be produced from photons alone, a wonderful realization of Einstein's equivalence of matter and energy. The mechanism for this process was proposed by **Patrick Blackett** (1897–1974) and **Guiseppe "Beppo" Occhialini** (1907–93).

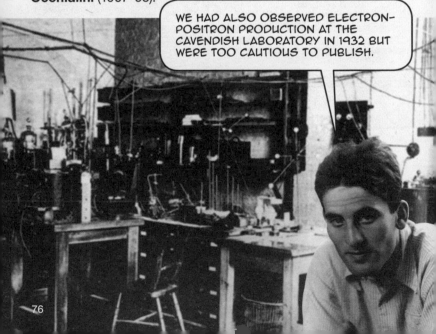

WE HAD ALSO OBSERVED ELECTRON-POSITRON PRODUCTION AT THE CAVENDISH LABORATORY IN 1932 BUT WERE TOO CAUTIOUS TO PUBLISH.

Blackett still won a Nobel Prize in Physics for his cloud chamber work. He developed a system in which the camera was controlled by a separate radiation counter such that the cosmic rays would take photographs of themselves.

THIS WAS THE FIRST EXAMPLE OF AUTOMATICALLY *TRIGGERING* A PARTICLE DETECTOR.

Further work allowed physicists to understand and make calculations relating to electromagnetic "showers" – the low-energy cosmic ray spray caused by cascades of photons producing electron–positron pairs, which would produce more photons, which would produce more electron–positron pairs … you get the idea.

Quantum field theory made some beautiful predictions that could be tested by experiment. What was particularly nice about it was that you could improve your calculations by taking into account lots of different processes that quantum mechanics tells you should *all* be taking place at the same time and affecting the final result.

What was particularly nasty about it was that this would often lead to the appearance of mathematical infinities in the final result. **Hans Bethe** (1906–2005) figured out a way around this while on a train in 1947 – he simply absorbed the troublesome infinities into measurable quantities that were known from experiment to be finite.

> THIS MATHEMATICAL TRICK CAME TO BE KNOWN AS *RENORMALIZATION*, AND IT FORMED THE BASIS OF THE INFINITE-FREE QUANTUM FIELD THEORY OF THE ELECTROMAGNETIC FORCE.

Quantum electrodynamics (QED) was first described in the literature by **Sin-Itiro Tomonaga** (1906–79), **Julian Schwinger** (1918–94) and **Richard P. Feynman** (1918–88).

> I HAD TO BE DIFFERENT, OF COURSE. NOT ONLY DID I USE A DIFFERENT MATHEMATICAL FORMULATION TO THE OTHER TWO GUYS, I WORKED OUT A WAY TO DO THE MATH USING PICTURES. MUCH MORE FUN!

Fun or not, "Feynman diagrams" correspond directly to the complicated algebra of QED. **Freeman Dyson** (b. 1923) showed that the formulations were equivalent. The calculations made possible by QED provide the most precise predictions testable by experiment, representing an utter triumph for theoretical and experimental physics: mind and matter in perfect harmony. However, some odd cosmic ray measurements nearly grounded the fledgling QED before it took off.

QED under threat

Calculations involving electromagnetic showers from cosmic rays could be tested against what was seen in cloud chambers. Carl Anderson had taken his cloud chamber up a mountain and was making such measurements with his student **Seth Neddermeyer** (1907–88) when they noticed curved tracks that didn't behave as they should.

AT FIRST THESE WERE THOUGHT TO REPRESENT A FAILURE OF THE NEW QUANTUM FIELD THEORY AT HIGH ENERGIES, BUT I FOUND TRACKS WITH ENERGIES WHERE THE THEORY HAD PREVIOUSLY WORKED WELL.

THE RESULTS COULD BE EXPLAINED BY THE EXISTENCE OF A NEW CHARGED PARTICLE WITH A MASS LARGER THAN THAT OF A NORMAL FREE ELECTRON BUT MUCH SMALLER THAN THAT OF A PROTON.

Rather than throw away the new field theory of electromagnetism, others made more precise measurements of this "penetrating radiation".

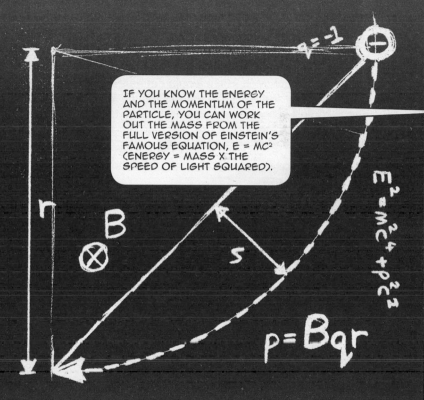

The mass was found to be roughly 130 times that of the electron (just over half of what it would later be measured as) – firmly between that of the electron and the proton. A new particle had been discovered. Funnily enough, quantum field theory had been used to show that it wasn't entirely unexpected … but things wouldn't quite be as straightforward as they were with the positron.

Mesons

While applying quantum field theory to the problem of what held the protons and neutrons together in the nucleus, theorist **Hideki Yukawa** (1907–81) suggested that a new particle should carry the nuclear force that prevented the positively-charged protons from flying apart. Like the photon carried the electro-magnetic force, this new particle would act as the carrier of this very strong nuclear force. Unlike the photon, it would have a mass as it could apparently act only across the tiny scales of the nucleus (the electromagnetic force, on the other hand, has an infinite range as the photon is massless).

I CALCULATED THE MASS SOMEWHERE BETWEEN THE ELECTRON AND THE PROTON, EVENTUALLY DUBBING IT THE *MESON* (HEISENBERG, WHOSE FATHER WAS A PROFESSOR OF GREEK, CORRECTED MY FIRST SUGGESTION OF "MESOTRON").

Yukawa's meson was an obvious candidate for Anderson and Neddermeyer's new particle – but that would have been too easy. Follow-up predictions by Tomonaga and others suggested that the meson should interact differently with ordinary matter depending on its charge. These predictions were confirmed by some cosmic-ray measurements but shown to be wrong by others. Bafflement ensued.

The solution was provided by a group at the University of Bristol led by **C.F. Powell** (1903–69). They had perfected a technique using plates of emulsion that would perform the role of both cloud chamber and photograph.

PARTICLES WOULD NOW TAKE AND DEVELOP THEIR OWN PICTURES.

Pions and muons

The emulsions not only allowed more precise track measurements; they were also much more portable. The particle-hunter no longer needed to drag a cloud chamber up to the ray-drenched mountain tops. Indeed, in 1947 Brazilian physicist and Bristol group member **César Lattes** (1924–2005) went to the 5,200m peak of Chacaltaya in the Andes with a set of emulsions. The results from this expedition and earlier missions in the Pyrenees identified two distinct particles: the charged **pion** and the **muon**.

$$105.7 \text{ MeV/c}^2$$

Yukawa was awarded the 1947 Nobel Prize in Physics, while Powell clinched the 1950 Prize. Anderson and Neddermeyer didn't get anything this time.

So while it was the pion that a (nuclear) quantum field theory had predicted, the muon had come out of nowhere. **Isidor Isaac Rabi** (1898–1988), discoverer of nuclear magnetic resonance, is famously said to have remarked of the muon:

WHO ORDERED THAT?

As we'll see later, we still aren't really sure.

The dawn of the age of accelerators

This wasn't the end of the pion story, though. Mathematical arguments based on symmetry had led **Nicholas Kemmer** (1911–98), a research fellow at Imperial College London, to suggest that there should also be an uncharged version of Yukawa's nuclear meson. Such a particle would be a good candidate for a source of photons in low-energy cosmic-ray showers: a neutral pion would decay into two photons.

HOWEVER, THEY WOULD BE TRICKY TO SPOT IN CLOUD CHAMBERS OR EMULSIONS AS NEITHER THEY NOR THEIR DAUGHTER PHOTONS WOULD IONIZE ANYTHING.

Finding evidence for the neutral pion would require a different approach that would fundamentally change the nature of particle physics and signal the beginning of the end of the particle-hunters' reign.

Naturally radioactive sources of particles were vital in the development of atomic and nuclear physics. However, they could only ever give an insight into naturally occurring matter. Cosmic rays, on the other hand, gave the patient particle-hunter access to the energies required to find new and exotic particles. Remember Einstein: more energy, more mass.

The trouble was that this placed the particle-hunters at the mercy of the cosmos. They had to take what the universe threw at them, on the universe's terms.

THE ODDS COULD BE IMPROVED BY GETTING A BIT CLOSER TO SPACE, BUT WHAT PHYSICISTS REALLY WANTED WAS CONTROL.

SO THEY SET ABOUT BUILDING MACHINES TO MAKE THEIR OWN COSMIC RAYS.

To do so was, in principle, simple. As we've seen, charged particles experience a force when placed in an electric field; this force accelerates them, giving them more energy. All that was needed was an electric field with a suitably high voltage. The first accelerators – Van de Graaff's famous generator, and **John Cockroft** (1897–1967) and **Ernest Walton**'s (1903–95) machine – could attain hundreds of thousands of volts.

AN *ELECTRON VOLT* IS THE ENERGY GAINED BY AN ELECTRON MOVED ACROSS A POTENTIAL DIFFERENCE OF ONE VOLT.

WHEN ACCELERATING PARTICLES, PHYSICISTS THEREFORE SPEAK OF ENERGY IN TERMS OF ELECTRON VOLTS – SYMBOL eV.

The Cavendish Laboratory's Cockroft–Walton machine was used in 1932 to perform the first artificial nuclear disintegration, winning the pair the 1951 Nobel Prize in Physics.

The Cyclotron and Synchrocyclotron

But the Cockroft–Walton machine was a one-shot wonder – accelerated particles would get only a single chance to feel the force. To achieve higher energies, **Ernest Lawrence** (1901–58) of the University of California, Berkeley, developed an accelerator in 1932 that used a magnetic field to bend the path of the accelerated particle round on itself: the Cyclotron.

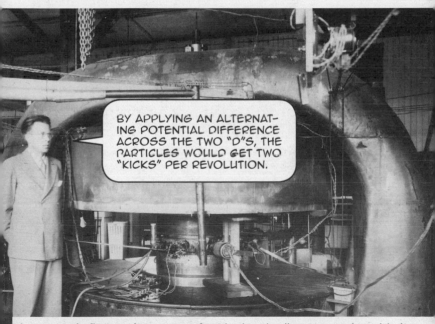

BY APPLYING AN ALTERNATING POTENTIAL DIFFERENCE ACROSS THE TWO "D"S, THE PARTICLES WOULD GET TWO "KICKS" PER REVOLUTION.

Lawrence's first cyclotron was four inches in diameter and could sit quite happily on the bench-top. They would get bigger. Much bigger.

It was actually a 184-inch Synchrocyclotron that was used at Berkeley to obtain indirect evidence for the neutral pion. It started operating in 1948 and could accelerate protons to a whopping 350 million electron volts (MeV).

MY SYNCHROCYCLOTRON ADJUSTED THE FREQUENCY OF THE ELECTRIC "KICKS" TO ACCOUNT FOR THE GAIN IN A PARTICLE'S MASS WITH AN INCREASE IN VELOCITY – A RELATIVISTIC EFFECT.

Edwin McMillan (1907–91)

In 1950 the machine was smashing protons into targets made of carbon and beryllium. The energy of the protons could be varied, and above the energy corresponding to the mass of the hypothesized neutral pion, photons were produced as expected.

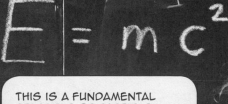

$$E = mc^2$$

THIS IS A FUNDAMENTAL PRINCIPLE OF PARTICLE PHYSICS: IN ORDER TO MAKE NEW PARTICLES, ONE NEEDS TO SUPPLY ENOUGH ENERGY TO MAKE THEIR MASS.

The Synchrotron

The particle hunters weren't done yet, though, and in the same year a group from Bristol used emulsions launched in balloons to observe the same process. Both results relied on spotting the photons converting to electron–positron pairs, which made seeing the requisite photon pair in the same pion decay difficult.

Of course, by this point the Berkeley group had built another, bigger accelerator. The electron Synchrotron could produce collimated* beams of X-rays at a specified energy from its electron beam. Synchrotrons varied the magnetic field of the accelerator to keep the electrons in a beam tube, rather than spiralling out from the centre of the synchrocyclotron. Results from this experiment showed the photons being produced in pairs in line with the neutral pion prediction.

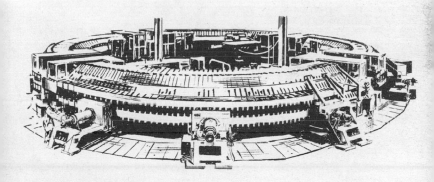

This discovery completed the meson picture, but more importantly it was the first particle to be discovered with an accelerator. It would the first of many.

In the time it took for the muon–pion confusion to be cleared up, there had been some strange developments on the cosmic ray front. During and just after the Second World War, there were five major cosmic ray groups, based at the École Polytechnique in Paris and the universities of Bristol, Manchester, Caltech and Berkeley. In 1943, **Louis Leprince-Ringuet** (1901–2000) of the École Polytechnique group, working in the French Alps, had spotted a particle with a mass three times that of the pion, but half that of the proton. In 1947, **George Rochester** (1908–2001) and **Clifford Butler** (1922–99) of the Manchester group found something just as strange.

WE TOOK TWO CLOUD CHAMBER PICTURES SHOWING DISTINCT FORKED TRACKS, OR V'S.

THESE SHOWED THAT THERE WERE HEAVY, UNSTABLE PARTICLES THAT DECAYED INTO THE NEWLY DISCOVERED PIONS.

They had opened a big can of particle-shaped worms. Similar events indicating unstable particles were found by the Bristol and Caltech teams; further new decays were found in 1951 when the Manchester group took their cloud chamber to the Pic-du-Midi in the Pyrenees. It took a grand meeting of the groups at Bagnères-de-Bigorre in July 1953, organized by Blackett and Leprince-Ringuet, to collate the results and present a coherent picture of the findings.

THIS CONFERENCE WAS FOR PARTICLE PHYSICS WHAT THE 1927 SOLVAY CONGRESS WAS TO QUANTUM PHYSICS.

But by coherently setting out the various decays, the particle-hunters had provided the accelerator physicists with an effective shopping list of decays and particles to look for.

Cosmotron, Bevatron and G-Stack

The 75-foot diameter, 3 billion electron-volt Cosmotron at the Brookhaven National Laboratory on Long Island quickly reproduced and expanded upon the conference results. As its name suggests, it had been purpose-built to make cosmic rays in the laboratory. In 1954, Berkeley hit back with the 180-foot diameter, 6 billion electron-volt Bevatron.

As these two collosi battled it out, the particle-hunters entered the fray with one spectacular "last hurrah": the G-Stack Collaboration of cosmic-ray physicists from across Europe organized a series of giant balloon flights exposing litres and litres of emulsion to rays at an altitude of up to 30 kilometres.

NOT ONLY DID WE PRODUCE RESULTS THAT MATCHED THOSE OF THE ACCELERATORS, THIS WAS THE FIRST COORDINATED INTERNATIONAL COLLABORATION BETWEEN INSTITUTIONS.

Strange times

Science relies on confirmation of results from independent sources, and the findings were consistent: something strange was going on with these new, unstable particles. Some of the decays predicted by the theories of the time were taking place as expected, but some weren't: nature was suppressing certain reactions, and making some of them occur far more slowly than theory predicted. **Murray Gell-Mann** (b. 1929), an American theorist, worked out that if you literally called some of the new particles "strange" (or, rather, assigned them the property "strangeness"*), the pattern and speed of the decays made sense.

THE DETAILS WILL MAKE MUCH MORE SENSE WHEN WE START TO TALK ABOUT THE STANDARD MODEL OF PARTICLE PHYSICS LATER ON.

Physicists hated the idea almost as much as the name "strangeness", but it stuck: the experimental results couldn't be argued with. Meanwhile, new particles were being discovered all the time. A breakthrough in detector technology helped with this: **Donald Glaser**'s (b. 1926) "bubble chamber", in which charged particles would leave photographable trails of bubbles in a super-heated liquid. While using a similar principle to that of the cloud chamber, it could make far better measurements of the short-lived strange decays.

> HOWEVER, YOU NEEDED TO KNOW EXACTLY WHEN TO EXPAND THE LIQUID – EASY IN ACCELERATOR EXPERIMENTS, NIGH-ON IMPOSSIBLE WITH COSMIC RAYS.

Glaser won the 1960 Nobel Prize in Physics for the invention of the bubble chamber, but it was the final nail in the coffin for the particle-hunters.

The particle zoo

The Bevatron went on to discover the **anti-proton** and the **anti-neutron**. The observation of *resonances* in production cross-sections – particles that exist for such a short amount of time (10^{-23} seconds) that the only way of inferring their existence is an increased amount of activity at the collision energy corresponding to their mass – saw a veritable zoo of new particles appear.

THE LIST OF KNOWN "ELEMENTARY" PARTICLES WAS GROWING SO LONG THAT THE FIELD WAS IN DANGER OF DEVELOPING AN ALMOST BIOLOGY-LIKE LEVEL OF COMPLEXITY.

Gell-Mann and others would go on to tame the zoo, but before we see how, we need to account for a great weakness in the tale told so far.

The neutrino

Earlier, we said that the proton, neutron, electron and photon provided us with all the particles needed for chemistry, biology, and all of the wonderful sciences that follow on from those. Actually, the atomic toolkit requires one more piece to describe all of the "everyday" phenomena, assuming you include radioactivity in the "everyday" category. We need the **neutrino**.

Rutherford's radioactive beta decay had been identified as the emission by the nucleus of an electron by Becquerel in 1900. However, later measurements of the ejected electron's energy left physicists scratching their heads: it varied continuously from electron to electron, suggesting that energy and momentum were not conserved in beta decay. This violated a fundamental tenet of physics.

IT WAS SO ODD THAT I SUGGESTED THAT ENERGY MIGHT ONLY BE CONSERVED *ON AVERAGE*.

Bohr

The weak force

Quantum hero **Wolfgang Pauli** (1900–58) stepped in with the requisite "this is odd, it's probably a new particle" hypothesis in 1930. **Enrico Fermi** (1901–54) seized on this hypothetical particle and the newly discovered neutron to develop a quantum field theory of beta decay. He named Pauli's light, undetectable particle the neutrino.

> IN MY FRAMEWORK, A NEUTRON CAN DECAY INTO A PROTON, AN ELECTRON AND A NEUTRINO.

The problem was that, compared to the other types of interaction in the nuclei, it was a very weak interaction. In typical imaginative fashion, physicists called this new force "the weak force".

Not only did the weak force explain radioactivity, it also helped out the cosmic-ray physicists. It was realized that charged pions were decaying, via the weak force, into muons and neutrinos. The neutrinos just couldn't be detected. Likewise, the longer-than-expected lifetimes observed in the strange decays made sense if one assumed a weak interaction was taking place, rather than a strong nuclear reaction.

The cosmic-ray physicists would return the favour: **Richard Dalitz**'s (1925–2006) summary of the tau–theta puzzle at the Bagnères-de-Bigorre meeting of 1953 would provide the motivation for a paradigm shift that would shake physics to its core.

THE TAU–THETA PUZZLE WAS THIS: THE *TAU* (AS IT WAS CALLED AT THE TIME) WAS A NEW, HEAVY PARTICLE THAT DECAYED TO THREE PIONS, WHILE THE *THETA* (AGAIN, ITS CONTEMPORARY NAME) DECAYED TO TWO.

Parity

However, the tau and the theta appeared to have the same mass, charge and decay lifetime. In particle terms, they were identical – except for one thing: their **parity**. A particle's parity can be thought of in terms of how it would look when reflected in a mirror. A spherical ball looks the same – it's invariant under a parity transformation. But a left-handed glove, for example, looks like a right-handed glove – it is said to change its parity.

> IT'S ACTUALLY MORE ACCURATE TO THINK ABOUT IT IN TERMS OF TURNING THE GLOVE INSIDE OUT.

Parity was thought to be a fundamental symmetry of nature – and why not? Why should the laws of physics be any different for left-handed things than for right-handed things?

If the tau and the theta were the same particle, that would be like saying a left-handed glove was magically becoming a right-handed glove: the two-pion and three-pion final states have different parities. If they were the same particle, it would mean that the symmetry did not hold and that parity was violated in our universe.

Two theoretical physicists, **Tsung Dao Lee** (b. 1926) and **Chen Ning Yang** (b. 1922), were inspired to examine all of the experimental evidence for parity conservation. They found that, amazingly, there wasn't any in the case of the weak interaction.

They suggested to their friend, **Chien-Shiung Wu** (1912–97), that she perform an experiment to directly test the violation of parity.

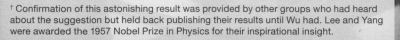

I USED THE BETA DECAY OF CRYOGENICALLY FROZEN COBALT-60 IN A MAGNETIC FIELD TO SHOW THAT THEY WERE RIGHT: PARITY WAS, IN FACT, VIOLATED IN WEAK INTERACTIONS†.

† Confirmation of this astonishing result was provided by other groups who had heard about the suggestion but held back publishing their results until Wu had. Lee and Yang were awarded the 1957 Nobel Prize in Physics for their inspirational insight.

Parity violation solved the tau–theta puzzle and led to the refinement of Fermi's theory of beta decay. But most astonishingly of all, it showed that the laws of physics *did* care if you were left-handed or right-handed (well, if you were a particle). A presumed fundamental symmetry of the universe had been shattered.

HOWEVER, FERMI'S THEORY DIDN'T EXPLAIN WHY THE WEAK FORCE WAS SO ... WEAK.

IT ALSO COULDN'T BE RENORMALIZED LIKE THE PHENOMENALLY SUCCESSFUL QED COULD.

There was clearly more work to do – work that would take over 50 years to complete, but would ultimately result in our current understanding of how matter and forces work together at the fundamental level: the **Standard Model** of particle physics.

The Standard Model

The Standard Model of particle physics is one of the crowning achievements of 20th-century science. One might therefore expect a more inspiring name, as with the elegant quantum electrodynamics (QED) or the evocative "Big Bang" theory of cosmology. Breaking it down into its component parts, though, there is some method to the mundanity.

"Standard" is straightforward enough to justify: its success in describing the interactions of matter has helped it to gain the widespread acceptance it enjoys among particle physicists. "Model", in theoretical physics, has a particular meaning: a model is the specific quantum field theory equation used to describe a given set of matter and forces.

THE STANDARD MODEL DESCRIBES IN EQUATION FORM EVERYTHING WE HAVE THUS FAR OBSERVED EXPERIMENTALLY, SO IT REALLY IS THE *STANDARD MODEL*.

The matter and forces are represented in the model by terms in this equation. There are terms for the matter particles – the **fermions** – which have an intrinsic quantum spin of one-half. There are terms for the force-carrier particles – the **bosons** – which have whole-number spin (photons, for example, have spin one). There are terms that represent the interactions between the matter and the force particles. There are nineteen physical constants that have to be measured from experimental data, and these determine properties like particle masses and the strengths of the forces (well, you have to give the experimentalists something to do).

IT'S IMPORTANT TO NOTE THAT THIS GARGANTUAN EQUATION WAS NOT WRITTEN DOWN IN ONE COFFEE-FUELLED BLACKBOARD BINGE. IT TOOK DECADES TO WORK IT ALL OUT.

The model has its limitations. It can't describe gravity – the first of the four fundamental forces of nature to be unlocked. Some would argue that nineteen arbitrary constants is eighteen too many for a truly beautiful "Theory of Everything". The final piece of the puzzle may only just have been found, but as we'll see at the end of the book, developments in neutrino physics mean we probably need to think about buying a whole new puzzle already. But, for now, the Standard Model really is our best Theory of Everything.

IT'S A BEAUTIFUL MATHEMATICAL JIGSAW THAT HAS BEEN CAREFULLY ASSEMBLED AND TESTED OVER THE COURSE OF HALF A CENTURY BY HUNDREDS – IF NOT THOUSANDS – OF PARTICLE PHYSICISTS.

Let's look at each part of the model as it's known today.

Quarks

We left the accelerator physicists struggling with the particle zoo of the 1950s. New particles just kept popping up everywhere. In his Nobel acceptance speech of 1955, **Willis Lamb** (1913–2008) opened with this zinger:

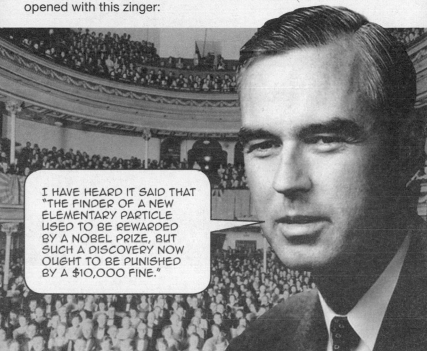

I HAVE HEARD IT SAID THAT "THE FINDER OF A NEW ELEMENTARY PARTICLE USED TO BE REWARDED BY A NOBEL PRIZE, BUT SUCH A DISCOVERY NOW OUGHT TO BE PUNISHED BY A $10,000 FINE."

Fortunately, Gell-Mann and (working independently) **Yuval Ne'eman** (1925–2006) had spotted another pattern in the particles. By arranging the zoo into groups according to their quantum mechanical spin, their strangeness, and their charge, a kind of order was restored. Gaps were spotted in these arrangements that pointed to undiscovered particles, which were subsequently found.

In 1962 Gell-Mann used the Eightfold Way* to predict the existence of the strange **Omega** particle. It was found in 1964 by the Brookhaven Alternating Gradient Synchrotron (AGS).

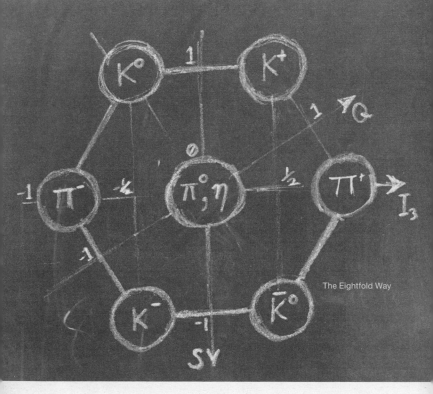

The Eightfold Way

By now accelerators were being built elsewhere, notably at the "Conseil Européen pour la Recherche Nucléaire" – CERN. Founded by twelve European governments in 1954, its aim was to harness the spirit of international collaboration to make European science competitive with that of America. Its first accelerator, a synchrocyclotron, was quickly succeeded by the 28 billion electron-volt Proton Synchrotron.

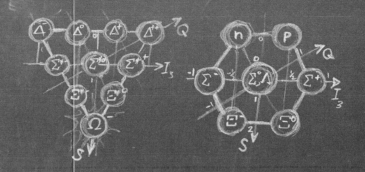

The Eightfold Way was the precursor to something more fundamental. Gell-Mann and (again, working independently) **George Zweig** (b. 1937), a former student of Feynman, realized that the organizing principle of the Eightfold Way – a symmetry – could be explained if the particles being discovered weren't elementary after all, but were themselves made up of smaller particles.

> I CALLED THEM *QUARKS*.

> I CALLED THEM *ACES*.

Gell-Mann's choice would win out.

109

Up, down, strange and charm

At first there were three quarks.

There were the "up" and "down" quarks. These were named after their quantum mechanical spin. A proton has two ups and a down, and a neutron has two downs and an up. There are the antimatter versions too: charged pions are made of an up and an anti-down, or a down and an anti-up.

And there was the "strange" quark. A particle's "strangeness" was then simply a count of how many strange quarks it contained! The strange quark had a greater mass than that of the up and the down, which is why the strange particles were heavier. The Omega **baryon*** was made up of three strange quarks (baryons contain three quarks or three anti-quarks).

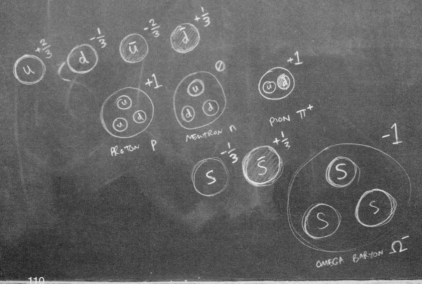

It didn't stop there. Work in 1964 by **Sheldon Glashow** (b. 1932) and **James Bjorken** (b. 1934), and later by Glashow, **John Iliopoulos** (b. 1940) and **Luciano Maiani** (b. 1941), showed that the existence of a fourth quark, the "charm", would explain why certain weak interactions weren't being observed. In 1974, competing teams at Brookhaven's AGS and the Stanford Linear Accelerator Center's (SLAC's) SPEAR electron-positron collider used higher accelerator energies to discover the **J/Psi** particle.

A new deluge of particles containing the even heavier charm quark were discovered in what has been called "the November Revolution".

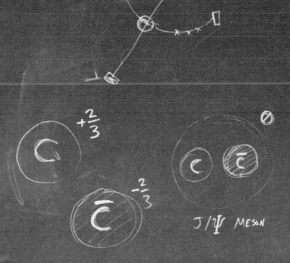

$+\frac{2}{3}$

$\emptyset$

$-\frac{2}{3}$

J/ψ MESON

Charge-parity

Would four quarks be enough? Apparently not. To see why, we need to return to the concept of parity. While physicists were shocked that parity was violated in our universe, they reassured themselves that the combination of parity with charge would form a new symmetry that was conserved – **charge-parity**, or CP for short.

A POSITIVE PARTICLE VIEWED IN THE "CHARGE" MIRROR WOULD APPEAR NEGATIVE – IT WOULD SEE ITS ANTIMATTER PARTNER.

JUST LIKE A LEFT-HANDED GLOVE WOULD LOOK LIKE A RIGHT-HANDED GLOVE IN THE "PARITY" MIRROR.

In the charge-parity mirror *both* charge and parity are switched. A universe that obeyed CP symmetry wouldn't care if you were left-handed or right-handed, as long as you had the correct charge too.

Physicists were happy with this until, in 1964, **James Cronin** (b. 1931) and **Val Logsden Fitch** (b. 1923) showed through their experiments with the neutral kaon system (kaons are like pions that have a strange quark) that the charge-parity symmetry was violated too![†] This left physics in bit of a mess: it was now known that the laws of physics were different for matter and antimatter.

Why this is the case is one of the greatest mysteries in physics, though the answer may go some way to explaining why we live in a universe made of matter and not antimatter.

[†] It is thought that when a third symmetry – the direction of time – is taken into account, the symmetry holds: the universe *is* Charge-Parity-Time, or CPT, invariant.

Bottom and top quarks

While the debate about *why* rumbled on, a mechanism explaining *how* CP violation could occur was proposed in 1973 by **Makato Kobayashi** (b. 1944) and **Toshihide Maskawa** (b. 1940), building on an idea of **Nicola Cabbibo** (1935–2010).

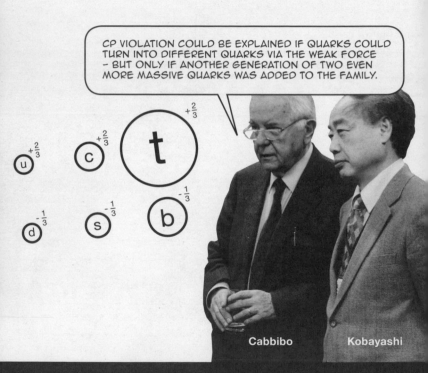

> CP VIOLATION COULD BE EXPLAINED IF QUARKS COULD TURN INTO DIFFERENT QUARKS VIA THE WEAK FORCE – BUT ONLY IF ANOTHER GENERATION OF TWO EVEN MORE MASSIVE QUARKS WAS ADDED TO THE FAMILY.

$u^{+\frac{2}{3}}$ $c^{+\frac{2}{3}}$ $t^{+\frac{2}{3}}$

$d^{-\frac{1}{3}}$ $s^{-\frac{1}{3}}$ $b^{-\frac{1}{3}}$

Cabbibo   Kobayashi

Driven by this prediction, the **bottom quark** was discovered by a team led by **Leon Lederman** (b. 1922) in 1977, while the extremely heavy **top quark** had to wait until 1995 until it was finally observed by two experiments at the Tevatron proton–anti-proton collider.

Both of these discoveries were made at America's rival to CERN, Fermilab near Chicago. It was founded in 1967 as the National Accelerator Laboratory (they added the "Fermi" in 1974 in Enrico's honour) by **Robert Wilson** (1914–2000).

Wilson, who had worked under (and was sacked twice by) Lawrence at Berkeley, was something of a free spirit – but he got things done, and done with style. He also refused to use military reasons to justify funding the new multi-million dollar laboratory. He told the Joint Congressional Committee on Atomic Energy:

IT HAS NOTHING TO DO DIRECTLY WITH DEFENDING OUR COUNTRY EXCEPT TO MAKE IT WORTH DEFENDING.

He also introduced a herd of bison to Fermilab. It's still there.

ROBERT RATHBUN WILSON H

Evidence that protons and neutrons were actually made of quarks, meanwhile, had come from **deep inelastic scattering** experiments at SLAC in 1968.

IN AN ECHO OF RUTHERFORD'S EXPERIMENTS OF 1911, ELECTRONS WERE FIRED AT PROTONS AND NEUTRONS AND WERE OBSERVED TO SCATTER IN A WAY CONSISTENT WITH A TARGET MADE UP OF THREE TINY POINT-LIKE PARTICLES.

The quark model wasn't really accepted at the time, so Feynman's term **parton** was used to describe these components of the nucleons – but further work showed that the quark model was indeed the right way to go.

So, to summarize:

- *There are six flavours of quarks, arranged in three generations.*

- *Each of these has a corresponding anti-quark.*

- *They have mass, and the mass of each flavour is heavier than the last.*

- *They have an electric charge and so experience the electromagnetic force.*

- *They also interact via the weak force, which is how they change from one flavour to another.*

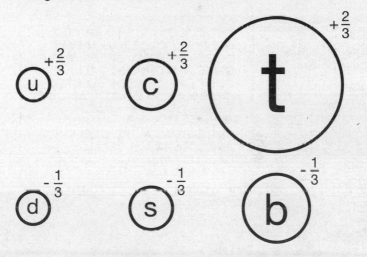

The zoo of particles is actually composed of different combinations of quarks and anti-quarks. These are the **hadrons**, which are split into two types: mesons* like the pion and the kaon are made of a quark–anti-quark pair; baryons* like the proton, the neutron, the Sigma and the Omega are made of three quarks or three anti-quarks.

The strong force and QCD

The strong force, one of the four fundamental forces of nature, holds the quarks (and anti-quarks) together. The corresponding quantum field theory of the strong force is **quantum chromo-dynamics** (QCD) and, like QED, it has its own force-carrying boson – in fact, it has eight of them – but it's a little more complicated than that.

For a particle to interact via the electromagnetic force, it has to have an electric charge. This allows it to swap photons with other charged particles. The equivalent of charge for the strong force is called "colour".

NOT LITERALLY DIFFER-ENT COLOURS, THOUGH: THE STRONG FORCE HAS THREE COLOURS (RED, GREEN AND BLUE), AS OPPOSED TO JUST ONE TYPE OF ELECTRIC CHARGE. YOU COULD CALL THE THREE TYPES OF CHARGE "+", "£" AND "!" IF YOU WANTED, BUT ACTUALLY THE COLOUR ANALOGY DOES BECOME USEFUL WHEN YOU START MIX-ING THEM TOGETHER ...

Gluons

There is actually only one type of electric charge, and the negative is just the anti-charge of the positive – so with QCD you have anti-red, anti-blue and anti-green too. Quarks have colour, and so can experience the strong force: they can swap **gluons**.

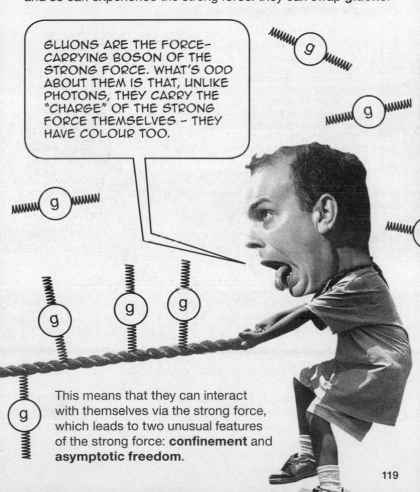

GLUONS ARE THE FORCE-CARRYING BOSON OF THE STRONG FORCE. WHAT'S ODD ABOUT THEM IS THAT, UNLIKE PHOTONS, THEY CARRY THE "CHARGE" OF THE STRONG FORCE THEMSELVES – THEY HAVE COLOUR TOO.

This means that they can interact with themselves via the strong force, which leads to two unusual features of the strong force: **confinement** and **asymptotic freedom**.

Confinement

Confinement justifies gluons' sticky name – the self-interactions mean that it's difficult to pull individual quarks apart. If you tried, it would take less energy to just create a new quark–anti-quark pair (a process called **hadronization**) – which is why you can never see individual quarks on their own.

THIS IS RATHER CONVENIENT, AS THE QUARKS HAVE FRACTIONAL CHARGE (TWO-THIRDS FOR THE UP-TYPES, ONE-THIRD FOR THE DOWN-TYPES) BUT THEY ALWAYS COMBINE IN A WAY THAT ENSURES THEY HAVE A WHOLE NUMBER CHARGE AND NO OVERALL COLOUR.

You either have a colour and its anti-colour – so they cancel out – or the three colours mix, like red, green and blue light combine to give white light. Don't take that too literally, though – it has nothing to do with photons!

Asymptotic freedom

Asymptotic freedom is a fancy way of saying that, at higher energies, the strong force paradoxically gets weaker. Again, this handily ensures that we can study QCD if we have a big enough accelerator. This feature of QCD was predicted independently by **David Politzer** (b. 1949) and **David Gross** (b. 1941) and his student **Frank Wilczek** (b. 1951), winning them the 2004 Nobel Prize in Physics. Evidence for gluons came from the PETRA experiment at DESY (Deutsches Elektronen-Synchrotron) in Germany.

ELECTRON–POSITRON COLLISIONS PRODUCED EVENTS WITH THREE PARTICLE "JETS", SPRAYS OF PARTICLES CAUSED BY THE HADRONIZATION OF TWO SINGLE QUARKS AND A SINGLE GLUON.

The leptons

fermi + boson
lept + quarks

Compared to the quarks and gluons, the **leptons** are much simpler beasts. Like the quarks, they are spin-half fermions. We have already met the electron and the muon, which belong to the first and second generation of the charged leptons. The third – the **tau lepton** – was found in experiments performed between 1974 and 1977 by **Martin Perl** (b. 1927) and co-workers using the SPEAR collider at SLAC.

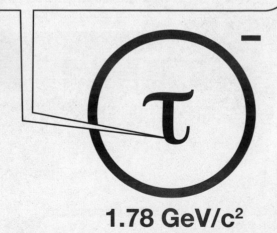

THE TAU WAS TRICKY TO FIND – IT HAS A MUCH GREATER MASS THAN ITS COUSINS AND SO DECAYS VERY QUICKLY AFTER IT'S CREATED – SO ITS EXISTENCE HAD TO BE INFERRED FROM VERY CAREFUL MEASUREMENTS OF WHAT WAS GOING MISSING IN THE DETECTOR.

1.78 GeV/c²

That said, in comparison to proving the existence of the neutrinos, the discovery of the tau was a walk in the park.

Twenty-six years passed between Pauli's neutrino hypothesis and the first positive results obtained by **Clyde Cowan Jr.** (1919–74) and **Frederick Reines** (1918–98).

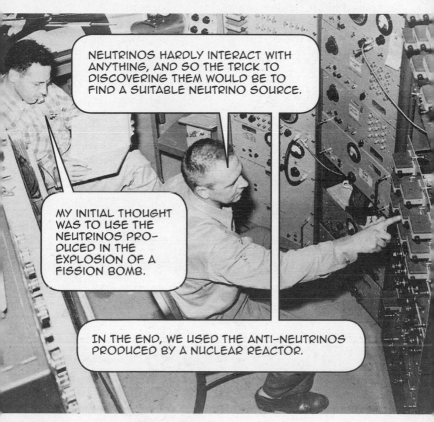

NEUTRINOS HARDLY INTERACT WITH ANYTHING, AND SO THE TRICK TO DISCOVERING THEM WOULD BE TO FIND A SUITABLE NEUTRINO SOURCE.

MY INITIAL THOUGHT WAS TO USE THE NEUTRINOS PRODUCED IN THE EXPLOSION OF A FISSION BOMB.

IN THE END, WE USED THE ANTI-NEUTRINOS PRODUCED BY A NUCLEAR REACTOR.

Anti-neutrinos from the reactor core were recorded interacting with protons in a tank of water, converting into a positron and a neutron that could be detected at the same time. Billions and billions of anti-neutrinos would pass through the detector every second, and yet only three or so events per hour would be usable.

Completing the lepton family

Cowan and Reines even made measurements with the reactor switched off. But switching a nuclear reactor on and off isn't a trivial task, and so physicists once again turned to accelerators to produce beams of neutrinos from the decay of pions.

In 1962, a team including **Melvin Schwarz** (1932–2006), **Jack Steinberger** (b. 1921) and Leon Lederman used such a beam produced by the Brookhaven AGS to show that there were actually two types of neutrinos.

ONE IS ASSOCIATED WITH THE ELECTRON AND ONE WITH THE MUON.

THE DISCOVERY OF THE **TAU NEUTRINO** IN 2000 BY THE DONUT COLLABORATION AT FERMILAB COMPLETED THE FAMILY OF LEPTONS: SIX FLAVOURS ARRANGED INTO THREE GENERATIONS, MIRRORING THE QUARKS.

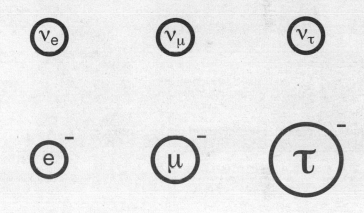

Leptons don't have colour, and so don't interact via the strong force. As you'd expect, the charged leptons interact via the electromagnetic force, and both families experience the weak force. What's more interesting is the masses of the leptons. The charged leptons have well-known masses, each generation heavier than the last. In the Standard Model, neutrinos have to be massless. More on that later …

Out of the zoo and into the pet shop

That completes the description of the "matter" particles in the Standard Model.

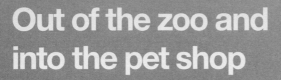

NOBODY KNOWS WHY THE SPIN-HALF FERMIONS ARE GROUPED SO NEATLY LIKE THIS, BUT THEY ARE.

IT'S CERTAINLY PREFERABLE TO THE ZOO, ALTHOUGH YOU COULD ARGUE THAT IT'S NOW MORE OF A "PARTICLE PET SHOP".

Now we turn to the physics at the heart of the Standard Model's beauty: the unification of the electromagnetic and weak forces, and how this ultimately ends up giving the universe the substance we know and love.

Electroweak unification

At first glance, the electromagnetic force and the weak force are pretty different animals.

ELECTRICITY, MAGNETISM, CHEMISTRY, BIOLOGY, ELECTRONICS, NOT FALLING THROUGH THE FLOOR, PARTY TRICKS INVOLVING BALLOON RUBBING – WE CAN (QUITE LITERALLY) SEE THE EFFECTS OF THE ELECTROMAGNETIC FORCE ALL AROUND US.

The weak force is far more subtle in its influence: radioactive decay quietly goes on all around us, but the mighty furnaces of the sun and her sisters all rely on this seemingly feeble, symmetry-violating interaction capable of transmuting some particles into other particles. Likewise, in the 1950s the mathematical theories used to describe each force were in very different states: the elegance and success of QED was somewhat at odds with Fermi's clumsier (though still useful) beta decay.

A new boson needed

The problem lay with the mechanics of the interaction: in Fermi's picture, four matter particles were meeting at a single point in space-time. This led to the dreaded spectre of infinities arising in the calculations.

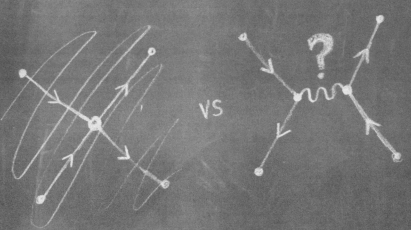

IF ONLY THE WEAK INTERACTION WOULD BEHAVE NICELY – IF ONLY IT HAD SOMETHING LIKE THE PHOTON BEING PASSED BETWEEN PAIRS OF PARTICLES, AS WITH THE ELECTRO-MAGNETIC FORCE – PHYSICISTS WOULD BE ABLE TO DO THE SUMS.

A new type of boson was therefore required to mediate the weak force. It would need to carry an electric charge, as it could change charged electrons into neutral neutrinos; and charged protons into neutral neutrons. It would also need to have its own mass to explain why the force was so weak.

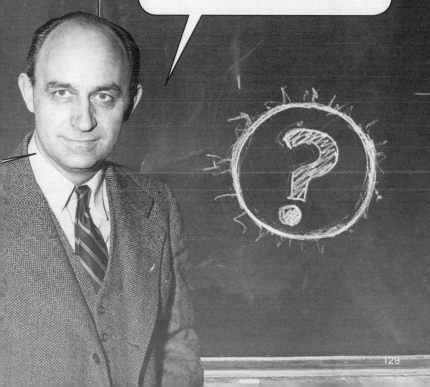

In 1954, Chen Ning Yang and **Robert L. Mills** (1927–99) worked out a theory that included three force-carrying bosons, each having an electric charge of –1, 0 and +1. This would handle the charge transfer of the boson exchange, but the three bosons still had to be massless.

The red herring was assuming that the neutral boson in the Yang–Mills theory was the photon.

But by 1968, **Abdus Salam** (1926–96), **Steven Weinberg** (b. 1933) and Sheldon Glashow had realized that a cunning combination of the massless photon and the massless Yang–Mills boson triplet would describe both the electromagnetic and weak forces in one fell swoop.

Salam and Weinberg, working at Imperial College London, used a technique called **spontaneous symmetry-breaking**.

WE SHOWED THAT THE FOUR MASSLESS BOSONS WERE ACTUALLY PART OF ONE FORCE, BUT COULD BE BROKEN UP INTO TWO FORCES IN SUCH A WAY THAT THREE (+1, 0, -1) BOSONS WERE LEFT WITH MASS WHILE ONE OF THE NEUTRAL BOSONS REMAINED MASSLESS.

Salam

It was Glashow who realized that the same massive bosons would be exchanged by the quarks, explaining the weak force in hadronic interactions.

131

Separated at birth

This unification of two fundamental forces was a massive achievement for both physics and reductionism. Just as Maxwell had unified electricity and magnetism, Salam, Weinberg and Glashow had shown that electromagnetism and the weak interaction were two forces separated at birth. In addition to photons, it was now supposed that there were massive W^+ and W^- bosons that were exchanged when charge changed hands between fermions (known as "charge current" interactions).

Annoyingly, there was now also this neutral, massive Z^0 boson for which no one had seen any evidence.

This wasn't surprising – these "neutral current" interactions didn't involve the exchange of any electric charge and so were difficult to spot – but it was only after **Gerardus 't Hooft** (b. 1946) and **Martinus Veltman** (b. 1931) showed that such theories were renormalizable that the search really began in earnest.

2006 Schertzer

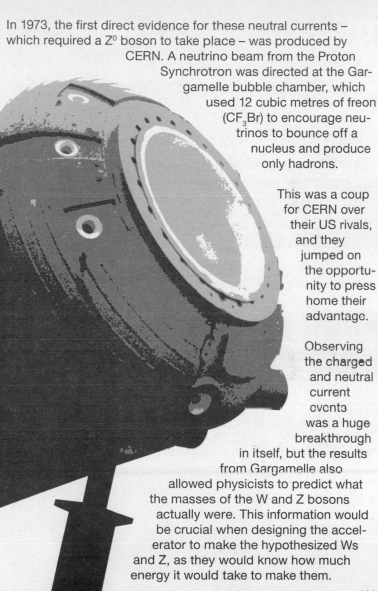

In 1973, the first direct evidence for these neutral currents – which required a Z^0 boson to take place – was produced by CERN. A neutrino beam from the Proton Synchrotron was directed at the Gargamelle bubble chamber, which used 12 cubic metres of freon (CF_3Br) to encourage neutrinos to bounce off a nucleus and produce only hadrons.

This was a coup for CERN over their US rivals, and they jumped on the opportunity to press home their advantage.

Observing the charged and neutral current events was a huge breakthrough in itself, but the results from Gargamelle also allowed physicists to predict what the masses of the W and Z bosons actually were. This information would be crucial when designing the accelerator to make the hypothesized Ws and Z, as they would know how much energy it would take to make them.

The two-beam collider

As it turned out, the energies were very large – nearly 100 giga electron volts (GeV), the equivalent of more than the mass of an atom of iron per boson. A new approach would be needed, and in 1976 a team of physicists including **Carlo Rubbia** (b. 1934) suggested that the energies required could be achieved by smashing beams of protons and anti-protons into each other head-on, rather than letting one beam plough into a fixed target. More energy meant more mass, and so more chance of making and discovering the bosons of the weak force.

SO IT WAS THAT CERN'S SUPER PROTON SYNCHROTRON (SPS) WAS CONVERTED INTO A TWO-BEAM COLLIDER: ONE FOR PROTONS AND ONE FOR ANTI-PROTONS.

This was made technologically possible by the work of **Simon van der Meer** (1925–2011), whose technique of "stochastic cooling" allowed enough anti-protons to be collected to make such collisions worthwhile.

The plan worked. In 1983, the two experiments situated at the points where the beams were brought together found the tell-tale signs of both the Ws and the Z^0.

> THIS DISCOVERY CEMENTED PHYSICISTS' BELIEF IN THE UNIFIED ELECTROWEAK FORCE AND THE STANDARD MODEL, WHICH WAS NOW COMING TOGETHER VERY NICELY INDEED.

The electroweak bosons had been discovered, but CERN didn't stop there. While Fermilab was building its Tevatron proton–antiproton collider – which as we have seen, still had a role to play with the discovery of the top quark – the CERN management saw that the next step was to study these new bosons in exquisite detail. To do that, they moved away from the messy hadron-on-hadron action of the SPS to the much cleaner collisions of electrons and positrons.

CERN's LEP collider: a crowning achievement

Matter and antimatter would be forced to annihilate at just the right energy to produce copious numbers of Zs and Ws. The Large Electron–Positron (LEP) collider, a 27km-circumference underground behemoth, was the boson factory that would enable physicists to test the Standard Model to the limit. From 1989–1995 it produced Z^0s, and from 1996–2000 it produced Ws. The precision measurements made with the four experiments dotted around the ring – ALEPH, DELPHI, L3 and OPAL – were a fitting tribute to the success of the 20th century's crowning physics achievement …

… A DESCRIPTION OF ALL KNOWN MATTER AND THREE OF THE FOUR FUNDAMENTAL FORCES CODIFIED IN THE EQUATIONS OF THE STANDARD MODEL.

The missing piece

Yet one piece of the Standard Model was missing, and LEP – and, indeed, Fermilab's Tevatron – hadn't been able to find it.

When Salam, Weinberg and Glashow unified the electromagnetic and weak forces, they did so by turning the four massless bosons of a unified electroweak force into three massive bosons – the W^+, W^- and Z^0 – and a massless boson – the photon. This in itself is a neat mathematical trick, but in discussing the unification of the forces earlier we naughtily glossed over the nub of the matter ...

... JUST HOW IT IS THAT THE Ws AND Z ACTUALLY GAIN THE MASS THAT MAKES THE WEAK FORCE SO, WELL, WEAK.

The problem of mass

Mass is one of those tricky concepts in physics that makes perfect sense right up until the point you start thinking about it. I mean *really* thinking about it.

$$F = -\frac{GMm}{r^2}$$

NEWTONIAN PHYSICS SAYS THAT MASS IS A PROPERTY OF AN OBJECT THAT TELLS YOU HOW MUCH FORCE YOU NEED TO APPLY TO ACCELERATE IT, OR THE FORCE EXERTED BETWEEN TWO OBJECTS DUE TO GRAVITATIONAL ATTRACTION.

SPECIAL RELATIVITY SAYS THAT IT'S THE AMOUNT OF ENERGY IN A BODY AT REST. GENERAL RELATIVITY* SAYS THAT IT'S A MEASURE OF HOW MUCH THAT BODY BENDS THE SURROUNDING SPACE-TIME.

$$E = mc^2$$

$$E^2 = m^2c^4 + p^2c^2$$

Those descriptions of what mass is might get more and more abstract – further and further away from our everyday experience – but they are all ultimately interpretations of a number in an equation.

Mass in a quantum field theory is no different. It's a number in an equation. It's another *m*. You can *try* to think of it as the energy associated with a quantum ripple in space-time when it isn't actually rippling (i.e. at "rest", whatever that means).

$$\mathcal{L}'_{EWK} = \mathcal{L}_{EWK} + \mathcal{L}_{\Phi \, kin.} + \mathcal{L}_{int.} + V(\Phi)$$
$$+ M_W^2 \, W_\mu^+ W^{-\mu} + M_Z^2 \, Z_\mu^0 Z^{0\mu}$$

BUT WHAT MATTERS IS THAT THE VALUE OF THE NUMBER AFFECTS THE CALCULATIONS IN A WAY THAT WE CAN MEASURE, AND MAKE PREDICTIONS ABOUT OTHER QUANTITIES THAT WE CAN TEST WITH EXPERIMENTS.

$$M_Z = \frac{M_W}{\cos \theta_W}$$

Conversely, the fact that experiments tell us that certain particles have mass means that we need to include these terms in the quantum field theory equations for them to be meaningful.

The problem is that we can't just stick these mass terms into the equations whenever we feel like it. This was the problem Yang and Mills had with their three-boson theory. If it's done in the wrong way, mass terms will lead to those pesky infinities appearing – i.e. the theory ceases to be renormalizable.

IN THE CASE OF THE ELECTROWEAK QUANTUM FIELD THEORY, SALAM AND WEINBERG JUST ASSUMED THAT IT WOULD ALL BE FINE; 'T HOOFT AND VELTMAN WON THE NOBEL PRIZE FOR SHOWING THAT THEIR ASSUMPTIONS WERE JUSTIFIED.

So how do you bequeath mass to the particles of the Standard Model? The key lay in the technique we mentioned earlier: sponta-neous symmetry-breaking.

The joys of symmetry

We've already noted how physicists love symmetry because, in general, it makes life easier. If the universe obeys a given symmetry, the laws of physics will be the same for everything in it with respect to that symmetry. If a symmetry of the universe is broken, we have to start worrying about physics being different for different things and life gets complicated.

In the early 1960s, **Yoichiro Nambu** (b. 1921) developed some ideas from another branch of physics – superconductivity – to find a way of breaking symmetries that didn't really destroy the underlying harmony of the equations.

In this way the symmetry wasn't fundamentally broken, as such – it was just hidden away.

Think of a marble in a wine bottle. If the marble is at the top of the "frog" (the rounded glass bump at the bottom), then the bottle and the marble will look the same from any direction (technically speaking, it exhibits a rotational symmetry about the vertical axis).

However, the system is unstable – the marble will spontaneously roll down into the trough running around the edge, breaking the symmetry.

The system has an underlying symmetry that is hidden – broken by the impracticalities of Nature (in this case, just how tricky it is to balance a marble on a rounded bump).

This is a pleasing notion for physicists: a hidden symmetry is better than no symmetry at all.

In 1961, **Jeffrey Goldstone** (b. 1933) showed that, in quantum field theories, spontaneous symmetry-breaking meant that new massless, spin-zero particles needed to be added to the equations. It was the existence of these new quantum fields that added the frogs to the wine bottle of the universe.

NOW COMES THE CLEVER BIT.

The Higgs mechanism

Nobody had seen any massless, spin-zero particles experimentally, so it was thought to be a nice trick but not much more than that. However, in 1963 **Philip Warren Anderson** (b. 1923) published a paper suggesting that these odd little spinless particles (called Nambu–Goldstone bosons) might be used to give spin-one bosons (like the Ws and Z) mass. Three groups of theorists independently seized upon the idea:

Peter Higgs
(b. 1929)

Robert Brout
(1928–2011)

François Englert
(b. 1932)

Gerald Guralnik
(b. 1936)

Carl Hagen
(b. 1937)

Tom Kibble
(b. 1932)

In what *Physics Review Letters* described as three of the "milestone" papers in its history, the summer of 1964 saw these three groups each publish papers about different aspects of the mechanism by which the massless spin-one bosons of Yang and Mills could gain a mass term as a result of the spontaneous symmetry-breaking trick.

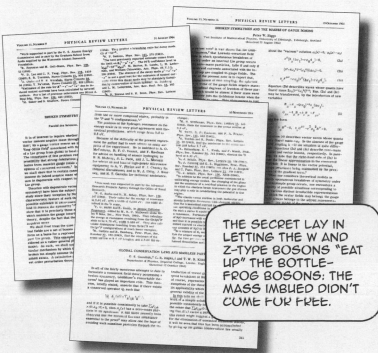

THE SECRET LAY IN LETTING THE W AND Z-TYPE BOSONS "EAT UP" THE BOTTLE-FROG BOSONS; THE MASS IMBUED DIDN'T COME FOR FREE.

It was actually Kibble who went on to show how the mechanism applied to quantum field theories involving Ws, Zs and photons. He also worked with Salam and Weinberg as they unified the electromagnetic and weak forces, using the Nambu–Goldstone bosons to give the Ws and Z mass while leaving the photon massless to unite the weak and electromagnetic forces.

Wanted: a massive, spin-zero boson

However, it was only Higgs who suggested that, after the Nambu–Goldstone bosons had been eaten by the massive bosons, there would be one spin-zero boson left that would actually have to have a mass itself. As it happened, the discovery of neutral currents, and of the Ws and Z themselves, seemed to validate the mass-generating mechanism used by Salam, Weinberg and Glashow in their scheme of electroweak unification.

However, for the Standard Model to be complete, this massive, spin-zero boson would need to be found experimentally.

The hunt for the **Higgs boson** was on.

The Higgs field

The fundamental problem with looking for the Higgs boson was
that the Standard Model, beautiful as it was, seemed to be very
coy about giving any indication of where its missing piece might
be found. In the Standard Model, the Higgs field permeates all
of space. Some particles interact with the Higgs field, and some
don't – it's all do with the mass terms in the equations.

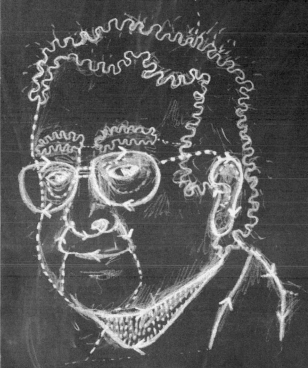

Particles with mass interact with the Higgs field, apparently slow-
ing them down. Massless particles (such as the photon) do not
interact with the Higgs field and so travel at the speed of light.

So how do you find evidence for this all-permeating Higgs field? Let's go back to our infinite number of quantum-mechanical springs. If enough energy is supplied by, say, a particle collision, real Higgs bosons can be created from ripples in the field. These Higgs bosons would then decay into particles you could measure. You just need enough energy to match the mass of the real Higgs boson. The trouble was, no one knew what this mass should be, and so how it might appear in an experiment. Theorist **Jonathan Ellis** (b. 1946) stated in a 1976 paper:

WE APOLOGIZE TO EXPERIMENTALISTS FOR HAVING NO IDEA WHAT IS THE MASS OF THE HIGGS BOSON ... FOR THESE REASONS WE DO NOT WANT TO ENCOURAGE BIG EXPERIMENTAL SEARCHES FOR [IT].

The measured ratio of charged to neutral currents had given experimentalists a mass estimate of the Ws and Z: this gave Rubbia et al. an energy target to hit with their CERN proton–anti-proton collider, which they went on to knock out of the park. The electroweak boson discoveries, and subsequent measurements of the Standard Model's properties at the LEP collider, lent more weight to the existence of this missing piece of the spontaneous symmetry-breaking puzzle.

A HIGGS BOSON WAS THE SIMPLEST EXPLANATION FOR THE BEAUTIFULLY MEASURED MASSES OF THE Ws AND THE Z, BUT IT WAS PAINFULLY MISSING FROM PARTICLE PHYSICISTS' TROPHY CABINET.

It was now time for some of those "big experimental searches".

The US strikes back

The US had lost a lot of ground to Europe, whose strategy of international cooperation through the establishment of CERN had reaped huge dividends. Work on the ISABELLE collider at Brookhaven, designed to find the W and Z, was halted in 1983: CERN had beaten the Americans to the electroweak discoveries, and ISABELLE's magnets weren't working anyway.

After a successful pitch to the Reagan administration, funds were reallocated to the collider to end all colliders.

The Superconducting Super Collider (SSC), an 87.1km-circumference machine lined with the superconducting magnets required to keep protons with 20 tera electron volts (TeV, 10^{12} electron volts) on track, would achieve collision energies far in excess of anything that had been achieved before.

Nicknamed the "Desertron", owing to its arid Texas location, it would find the Higgs and bring the United States back to the forefront of fundamental physics.

Or at least that was the plan. The barely-concealed nationalism surrounding the project was a big turn-off for potential partners from overseas, whose help might have offset the super-budget that the super-collider required.

The Large Hadron Collider

In the meantime, Rubbia was seeking such support for CERN's own Higgs hunter: the Large Hadron Collider, affectionately known in scientific circles as the LHC. Though "only" aiming for 7 TeV per beam, the LHC had an immediate advantage over its American rival – the 27km-long tunnel had already been built for the LEP collider, which it would replace. The financial viability of recycling CERN's civil engineering efforts – along with their recent success with the Ws and Zs – made the LHC an attractive proposition for international collaborators.

THE TRANSATLANTIC TITLE FIGHT WAS SHAPING UP TO FEATURE THE U.S. ALONE IN THE RED CORNER, WHILE IN THE BLUE CORNER THE REST OF THE WORLD LINED UP BEHIND CERN.

Sadly, the contest was over before it began. Plagued by an ever-increasing budget, political opposition and competition from the International Space Station, the SSC was officially cancelled on 21 October 1993 – but not before $2 billion had been spent and fourteen miles of the tunnel had been excavated in the Texan desert.

> AS IF TO KICK SAND IN THEIR RIVALS' FACES, THE FINAL LHC DESIGN STUDY WAS PUBLISHED JUST A MONTH LATER.

The project was formally approved by the CERN council in December 1994. The LHC would be going it alone in its hunt for the Higgs, and so began a mass migration of scientists from the SSC to the LHC.

Tension at CERN

Of course, this all hinged on the assumption that the Higgs boson wouldn't be found first elsewhere. The US still had the Tevatron, and the LEP collider itself still had the potential to find a sub-100 GeV Higgs boson. As it happened, this became a major source of tension at CERN. In the late 1990s, the LEP collider engineers did extraordinary work to keep pushing up the collision energies.

BUT WITH THE LEP STILL RUNNING, THE LHC ENGINEERS COULDN'T GET STARTED.

The LEP collider was finally switched off on 2 November 2000, and work began on the LHC a month later. In fact, some scientists claimed to have seen hints of the Higgs boson in their experiments at the LEP collider, making the CERN management's decision even more painful for all concerned.

AS IT TURNED OUT, THEY MADE THE RIGHT CHOICE – BUT THEN HINDSIGHT IS A WONDERFUL THING.

So how would the LHC achieve its goal of finding the Higgs boson? The rule was as simple as it had ever been: focus as much energy in as small a space as possible and hope that enough Higgs bosons would pop into existence long enough for the watching physicists to spot them.

155

A new frontier

The energy would be provided by two colliding beams of protons. These protons would be injected into the 27km-circumference accelerator ring in both directions – clockwise and anticlockwise – from the W and Z-discovering SPS accelerator.

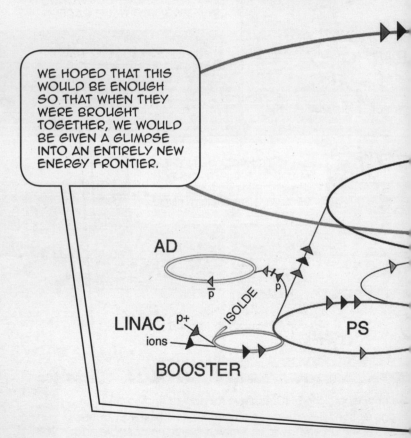

WE HOPED THAT THIS WOULD BE ENOUGH SO THAT WHEN THEY WERE BROUGHT TOGETHER, WE WOULD BE GIVEN A GLIMPSE INTO AN ENTIRELY NEW ENERGY FRONTIER.

One of the biggest problems the LHC engineers had to solve was how to keep protons with these energies – some 99.9999991% of the speed of light – on the right track.

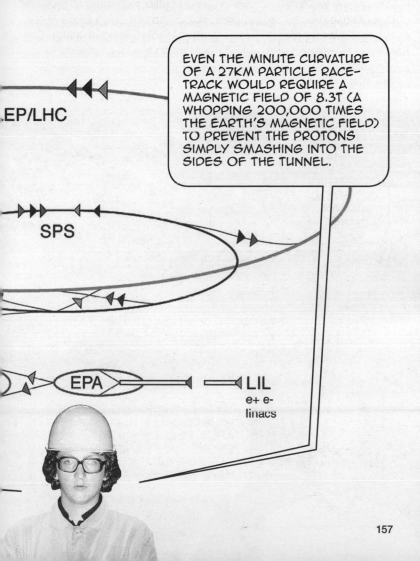

EVEN THE MINUTE CURVATURE OF A 27KM PARTICLE RACE-TRACK WOULD REQUIRE A MAGNETIC FIELD OF 8.3T (A WHOPPING 200,000 TIMES THE EARTH'S MAGNETIC FIELD) TO PREVENT THE PROTONS SIMPLY SMASHING INTO THE SIDES OF THE TUNNEL.

The answer lay in superconducting magnet technology. By cooling the magnets to a temperature of 1.9K (−271°C), the coils of the electro-magnets would essentially have zero electrical resistance, making it possible to produce the large currents required to generate the beam-bending magnetic fields. 1,232 15m-long dipole magnets would bend the beams and 392 7m-long quadrupole magnets would keep them focused in the beam pipes. The beam pipes themselves would be kept at a vacuum of 10^{-13} atmospheres (less than that of outer space), and operated at a temperature a few degrees above absolute zero (again, less than that of outer space).

TO FIND THE HIGGS BOSON, WE BASICALLY HAD TO CREATE A RING OF OUTER SPACE UNDER THE FRANCO-SWISS BORDER. CROOKES WOULD HAVE BEEN PLEASED.

Life would be further complicated by the need to "cross the streams" in order to actually collide the protons together. Another type of magnet would be used at four points on the LHC ring to bring the hair's-width beams crashing together – a feat that could be compared to colliding a pair of knitting needles fired from either side of the Atlantic.

At these "interaction points" the quarks and gluons of the hapless protons would interact, and – should they exist, and if enough energy to create them was available from the collision – new particles would be born of these fleeting quantum-mechanical liaisons.

From there on, the accelerator physicists and engineers could do no more: it would be down to the experimental physicists to prove that the new particles existed. Four experiments at four interaction points were proposed, designed and built in parallel to the LHC. And, much like the accelerator itself, these would be of a scale never before seen in the history of particle physics.

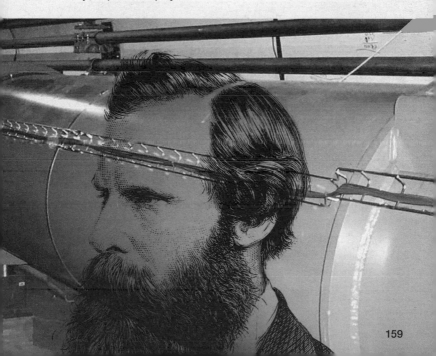

Two big challenges

Two challenges were presented by the hunt for the Higgs boson – or whatever phenomenon was responsible for electroweak symmetry-breaking. Firstly, physicists didn't know the expected mass of the Higgs boson, and so didn't know what they were looking for. Its mass would determine which of the more familiar particles the Higgs boson would decay into, and so which particles the experimenters would actually see in their detectors.

THE DETECTORS WOULD THEREFORE HAVE TO BE DESIGNED TO MEASURE AS WIDE A RANGE OF EXPERIMENTAL SIGNALS AS POSSIBLE.

To do this, equipment was placed around the interaction points in concentric layers with each layer performing a different function.

The inner layers were typically for measuring the curvature of charged-particle tracks, giving a measurement of the particles' momentum. Further out, the calorimetry systems would measure the energy deposited by particles churned out by the proton–proton collisions. Muons, the heavier cousin of the electron, needed their own detectors further out still, as they tended to punch through the tracker and calorimeter apparatus.

Some experiments would use Čerenkov detectors, making use of the fact that the speed of light in certain materials is slower than it is in a vacuum.

Particles travelling faster than this slower speed create an electromagnetic "sonic boom" that enables a measurement of the particle's velocity to be made.

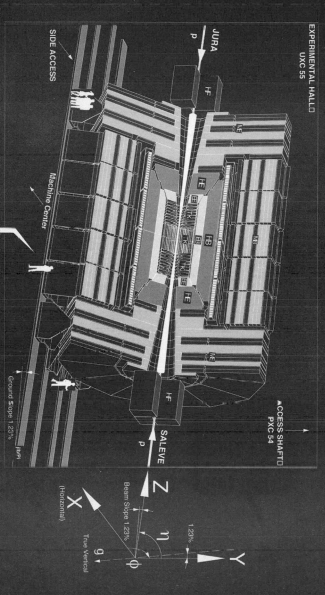

The CMS Detector at point 5 of LHC

© 1997 CERN/CMS Collaboration

161

The combination of velocity, momentum and energy measurements, as well as the pin-pointing of secondary particle vertices, would allow physicists to reconstruct and identify the particles made in the high-energy collisions. New particles could not be photographed directly: rather, the particle-hunters would be piecing together information about their quarry from what they left behind – droppings, tracks, disturbances in the undergrowth.

AND THE MORE EXOTIC THE PREY, THE MORE DIFFICULT THE HUNT BECAME.

The second challenge compounded that of the first – the Higgs boson was expected to be produced very rarely. Each potential Higgs sighting was a one-in-several-billion shot. The LHC was therefore built to collide billions and billions of protons together every second in order to maximize the chances of discovery – but to obtain the evidence of such a discovery the experimentalists would have to sift through billions of potential Higgs boson signals every single second.

Improvements in particle detectors

All particle detection relies upon converting the ionization caused by charged particles into useful information. Hans Geiger and Walther Müller's eponymous tube used the ionization of an inert gas to create an electrical signal that could be used to count the number of charged particles that entered the tube: you probably know this as the Geiger counter.

BUT THE INFORMATION PROVIDED BY MY DETECTOR WAS FAIRLY LIMITED - IN A GIVEN TIME WINDOW, YOU KNEW ONLY IF A CHARGED PARTICLE HAD BEEN PRESENT, OR IF IT HADN'T.

No measure of the particle's energy could be made, and the only spatial information available came from the position of the detector itself. The cloud chamber (and its successor the bubble chamber) and photographic emulsions, on the other hand, offered physicists a wealth of information in the form of the tracks of ionized droplets (or bubbles) that could be photographed, or the trails of grains from nuclear emulsions.

But this information came at a cost: photographs and emulsions needed to be developed and the results had to be processed and analysed by eye, all of which took time and human effort. Furthermore, the nature of the interacting media meant that there was a limit to the number of events could be recorded in a given amount of time – for example, bubble chambers would need around a second or two between bombardments in order to differentiate between separate particle collisions.

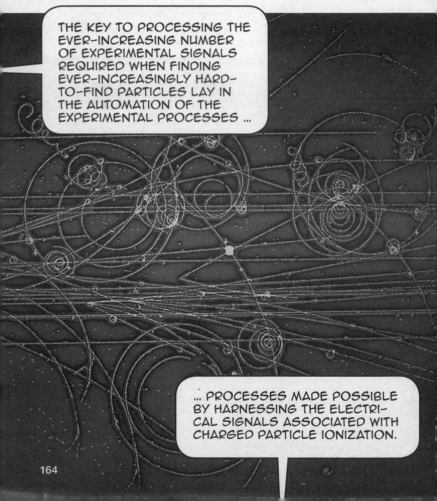

THE KEY TO PROCESSING THE EVER-INCREASING NUMBER OF EXPERIMENTAL SIGNALS REQUIRED WHEN FINDING EVER-INCREASINGLY HARD-TO-FIND PARTICLES LAY IN THE AUTOMATION OF THE EXPERIMENTAL PROCESSES ...

... PROCESSES MADE POSSIBLE BY HARNESSING THE ELECTRI-CAL SIGNALS ASSOCIATED WITH CHARGED PARTICLE IONIZATION.

Patrick Blackett had already won a Nobel prize for using Geiger–Müller tubes to trigger a cloud chamber's camera when it detected the presence of a high-energy charged particle. The **spark chamber** used charged metallic plates arranged within a volume of inert gas to create electrical sparks where the charged particles had created ions – a kind of electric cloud chamber that provided spatial information in a rapid-fire electrical form.

A great step forward was the **multiwire proportional chamber**, invented by **Georges Charpak** (1924–2010) at CERN in 1968. By dividing the plates into wires of metal running through the detector, further spatial information could be obtained from the electrical signals.

ARRANGING PLANES OF WIRES IN COMPLEMENTARY ORIENTATIONS ALLOWED A THREE-DIMENSIONAL PICTURE OF THE CHARGED-PARTICLE TRAJECTORIES TO BE ASSEMBLED.

The arrival of computers

But the wires offered something far more useful than positional granularity: the number of ions produced in the gas of the chamber was proportional to the energy of the incident particle. As the ions hit the wires, propelled by the high voltages applied to the wires themselves, they created an electric current that gave a measurement of the particle energy: more energy, more ions, more current. This signal could then be read out by the detector electronics and – crucially – stored on magnetic tape for later analysis. **Drift chambers** further developed the idea, using the timing of the pulses to refine the measurements made.

```
// Constants
//
static const uint16_t FEDCH_PER_DELAY_CHIP = 4;
static const uint16_t DELAY_CHIPS_PER_FED = FEDCH_PER_FED/FEDCH_PE
static const uint16_t SPY_DELAY_CHIP_PAYLOAD_SIZE_IN_BYTES = 376*4
static const uint16_t SPY_DELAY_CHIP_BUFFER_SIZE_IN_BYTES = SPY_DI
static const uint16_t SPY_DELAYCHIP_DATA_OFFSET_IN_BITS = 44;
//static const uint16_t SPY_SAMPLES_PER_CHANNEL = ( (SPY_DELAY_CH
// TW Dirty hack to lose the 3 samples from the end that screw th
static const uint16_t SPY_SAMPLES_PER_CHANNEL = 298;
static const uint16_t SPY_BUFFER_SIZE_IN_BYTES = SPY_DELAY_CHIP_
// Delaychip data + 8 bytes header for counters + 8 bytes for wo
// + 16 bytes for DA       der and trailer

//
// Class definitio

//
//class represent
class FEDSpyBuffe
{
public:
    //construct from buffer
    FEDSpyBuffer(const uint8_t* fedBuffer, const size_t fedBuff
    virtual ~FEDSpyBuffer();
    virtual void print(std::ostream& os) const;

    //get the run number from the corresponding global run
    uint32_t globalRunNumber() const;
    //get the L1 ID stored in the spy header
    uint32_t spyHeaderL1ID() const;
    //get the total frame count stored in the spy header
    uint32_t spyHeaderTotalEventCount() const;
    //get the L1 ID after reading a given delay chip
    //get the L1 ID after reading a given delay chip
    // ayChipL1ID(const uint8_t delayChip) const;
    //           nt after reading a given delay
    //           nt after reading a given uint8_t delayCh
```

> THE AUTOMATION OF PARTICLE MEASUREMENT AND IDENTIFICATION WITH COMPUTER CODE COULD NOW BEGIN.

The ultimate (at least, for now) in high-precision, high-speed detector technology essentially brought the computer inside the particle detector: slices of silicon semiconductor fitted to sophisticated microprocessors could provide micron-level positional measurements every couple of nanoseconds. These **silicon detectors** could function as the electronic bubble chambers required to cope with the collision rates of the LHC.

s.h"

```
// 376 32bit words
// Extra 8 bytes for c
// Offset to start of
ELAY_CHIP;                              YCHIP DATA_OFFSET_IN_BITS )  10
Y_CHIP_PAYLOAD_SIZE_IN_BYTES+8;

BUFFER_SIZE_IN
is up...

FER_SIZE_IN_BY
with delay chi
```

IRONICALLY, THESE DETECTORS WOULD ACTUALLY PRODUCE TOO MUCH INFORMATION EVEN FOR COMPUTERS TO ANALYSE IN A SENSIBLE AMOUNT OF TIME.

```
ize);
```

The selection of potentially interesting events would require some Blackett-style triggering mechanism.

167

```
ip
    const;
me from the same event
```

<h1>Information man-agement: the internet and the Grid</h1>

<p>The triggering systems of the LHC experiments were designed to be detectors-within-detectors: relatively crude measurements that might indicate the presence of something like a Higgs boson were made by sub-systems operating in tandem with the mother experiment. An interesting signal from the trigger would fire off the readout and storage of information from the whole experiment. Even then, the analysis of triggered data churned out by a typical particle physics experiment would require new mechanisms for sharing and processing detector data.</p>

<p>In 1991, **Tim Berners-Lee** (b. 1955) made the first website at CERN, brilliantly joining up the various internet technologies available at the time to create an information management system known as the World Wide Web.</p>

<p>The LHC would need more than this, however: the distributed information-processing capabilities offered by the "Grid" would be needed to turn the terabytes of detector output every day into useful physics results. Particle physics had come a long way from Anderson's painstaking analysis of 1,300 hand-triggered cloud chamber photographs to find those few glimpses of anti-matter.</p>

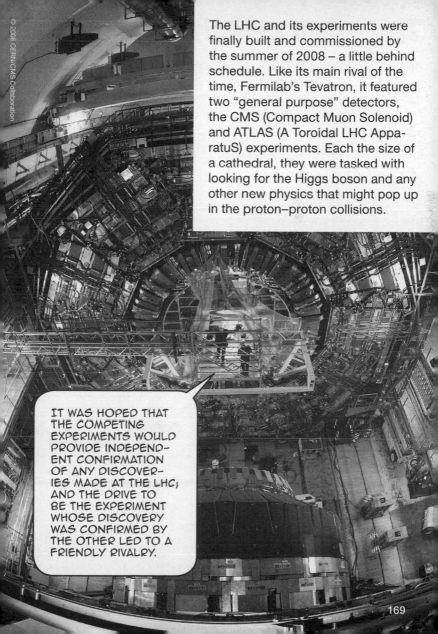

© 2008 CERN/CMS Collaboration

The LHC and its experiments were finally built and commissioned by the summer of 2008 – a little behind schedule. Like its main rival of the time, Fermilab's Tevatron, it featured two "general purpose" detectors, the CMS (Compact Muon Solenoid) and ATLAS (A Toroidal LHC Apparatus) experiments. Each the size of a cathedral, they were tasked with looking for the Higgs boson and any other new physics that might pop up in the proton–proton collisions.

IT WAS HOPED THAT THE COMPETING EXPERIMENTS WOULD PROVIDE INDEPENDENT CONFIRMATION OF ANY DISCOVERIES MADE AT THE LHC; AND THE DRIVE TO BE THE EXPERIMENT WHOSE DISCOVERY WAS CONFIRMED BY THE OTHER LED TO A FRIENDLY RIVALRY.

The great switch-on

On 10 September 2008, the world's media gathered in Geneva for the biggest switch-on in the history of particle physics. Finally, the largest and most complex machine humanity had endeavoured to build was to start running. The answer to the ultimate question – "What gives particles mass in the Standard Model?" – was due to be answered.

In the CERN Control Centre, LHC project mastermind **Lyn Evans** (b. 1945) led the countdown.

Many journalists were bemused by the raucous cheers and applause that greeted the tiny flash on mission control's wall-mounted screens.

But those flashes, proving to the assembled crowd that the beam of protons had successfully navigated all 27km of the LHC's accelerator ring, represented decades of research, ingenuity and hard work that would now lead to the answer to the ultimate question.

OR AT LEAST IT WOULD HAVE, IF THE LHC HADN'T BROKEN DOWN NINE DAYS LATER.

A faulty electrical connection between two of the superconducting magnets caused a spark to puncture the helium enclosure, initiating multiple magnet quenches and the leaking of two tonnes of helium into the underground tunnel. Fortunately no one was hurt, but around 50 of the LHC's magnets were damaged and had to be replaced in the subsequent shutdown.

But thanks to tireless work from CERN's accelerator engineers, the LHC was back up and running by November 2009 and achieved a world-record collision energy of 7 TeV in March 2010. While this still opened a new energy frontier for physicists, 3.5 TeV per beam was only half the design energy – the magnets would require further work to take the proton beams to what had been intended. The question now was: would 7 TeV be enough to find the Higgs boson?

The early days – when any shockingly new physics would have made itself known – passed largely without incident.

THE LACK OF NEW AND EXCITING PARTICLES CAUSED SOME TO WONDER ABOUT THE CONSEQUENCES OF THE LHC FINDING NOTHING AT ALL.

Another challenge from the US

In the meantime, the Tevatron was busy trying to pip the LHC to the post. It didn't have the energy or the collision rate, but it was still able to rule out Higgs bosons in certain mass ranges as time pressed on. Could the tortoise scoop the hare?

As it happened, it did not. After two years of running, CERN took the unusual step of announcing results as they went along, presumably to combat previous problems with "rogue" physicists releasing unchecked results from the Tevatron and the LHC on weblogs.

IN DECEMBER 2011, "TANTALIZING HINTS" OF A HIGGS-LIKE SIGNAL WERE APPEARING IN THE DATA.

On 4 July 2012, ATLAS spokesperson **Fabiola Gianotti** (b. 1962) and her CMS counterpart announced that a particle with a mass of 125 GeV had been discovered, consistent with the boson proposed by Higgs nearly 50 years previously. As CERN's director general put it in the following press conference:

WE HAVE IT.

Rolf Heuer

It finally looked like the Standard Model was complete. The origin of mass, the spontaneous symmetry-breaker, the missing piece of the particle physics jigsaw puzzle – whatever you want to call it – had been found. But while the LHC and its experiments had triumphed in its primary objective, many questions remained. One journey was over, but where could particle physics go now?

Unfinished Higgsness

Given the hype, it's very hard to believe that what was found in 2012 was *not* the last piece of the Standard Model. Further data will be required to verify that it is, in fact, the first spin-zero particle ever to be observed (all others have been spin one-half or spin-one). However, even a Standard Model Higgs boson isn't without problems, and it certainly doesn't end the story of particle physics.

SHARPER READERS MAY HAVE NOTICED SOMETHING A LITTLE ODD: THE BOSON THAT GIVES OTHER PARTICLES MASS *HAS A MASS ITSELF*.

This isn't a fundamental problem in terms of the quantum field theory – the mathematics just about makes sense – but odd things happen when you try to calculate what the mass of the Higgs boson should be.

As we saw with quantum electrodynamics, measurable numbers like the mass of a particle are subject to quantum corrections arising from loops in the Feynman diagrams. In the case of the Higgs boson mass, extra terms in the equation come from every type of massive particle thought to exist. While these don't lead to intractable infinities, they do lead to very, very large numbers (a 1 followed by 30 zeros) that have to be "fine tuned" away to give the Higgs mass observed in experiment. This is known as the **hierarchy problem**, and while it doesn't mean that we have to throw the Higgs boson out, physicists are uncomfortable with sticking in absurdly large numbers where they're not really wanted.

1000000000000000000000000000000

SO WHILE THE HIGGS BOSON IS THE LAST PIECE OF THE STANDARD MODEL PUZZLE, MOST PHYSICISTS HOPE THAT THE STANDARD MODEL ISN'T THE LAST PUZZLE IN THE TOYSHOP.

Supersymmetry and dark matter

So what could follow the Standard Model? One of the big hopes for a "new physics" discovery at the LHC was a theory called **supersymmetry**. Developed in the 1970s, supersymmetry is a symmetry that relates matter particles – fermions – to force-exchange particles – bosons. Just as matter and antimatter are linked by Dirac's mathematical trickery, so too were these fundamentally different components of reality. Every type of particle would have a supersymmetric particle – or **sparticle** – as a partner.

ASIDE FROM THE AESTHETIC JUSTICE THIS WOULD BRING TO THE UNIVERSE - AFTER ALL, WHY SHOULD MATTER AND FORCES BE DIFFERENT THINGS? - THESE PARTNER PARTICLES SOLVED THE HIGGS BOSON'S HIERARCHY PROBLEM.

The sparticle loops cancelled out the embarrassingly large numbers in the Higgs mass correction equations.

Of course, the number of fundamental particles instantly doubles when you introduce supersymmetry to your universe (in fact, you also end up with not one but *five* Higgs bosons).

BUT THIS IS SEEN AS A PRICE WORTH PAYING, ESPECIALLY IF ANY EXPERIMENTAL PROOF FOR SUPERSYMMETRY WERE ACTUALLY FOUND.

Such proof has so far eluded the physicists of the LHC. The tell-tale signal would be very massive particles disappearing from the proton–proton collisions after their creation: sparticles are predicted to be invisible to the LHC's detectors, and so would be spotted by looking for stuff systematically going missing. Incidentally, this is what makes supersymmetry a potential explanation for the mysterious **dark matter**, the invisible stuff that cosmologists believe makes up just over a quarter of the known universe.

The graviton

Supersymmetry also provides a way of neatly unifying the three fundamental forces of the Standard Model at a single (but very large) energy – the extra sparticles again helping out with the quantum corrections. And it suggests a candidate particle for the **graviton** – the hypothetical exchange boson of gravity, the fundamental force that most Standard Models just can't reach. So while there's no experimental evidence for super-symmetry, there are lots of convincing reasons to keep looking.

> GRAVITONS APPEAR IN SUPERSYMMETRY BECAUSE THEY HAVE AN INTRINSIC SPIN OF TWO, AND ARE REQUIRED FOR THE SUPERSYMMETRIC MATHEMATICS TO WORK.

Unfortunately, spin-two particles can't be worked into renormal-izable quantum field theories – which is why there's no theory of quantum gravity yet – but this hasn't stopped experimentalists looking for gravitons in other ways at the LHC.

Gravity, extra dimensions and (micro-) black holes

Compared to the other three forces, even the so-called "weak" force, gravity is pathetic. It only attracts things, and you need planet-sized objects for it to make any noticeable difference to your life. But why is this the case? Though there's no evidence for the graviton, particle physicists are happy to put Einstein's warping of space-time interpretation to one side and assume that some sort of boson is responsible. One way of looking for gravitons at the LHC would then be to see if they were disappearing upon being created in the high-energy proton–proton collisions. However, there is another way.

SOME MATHEMATICAL THEORIES – LIKE **STRING THEORY** – SUGGEST THAT OUR UNIVERSE IS COMPOSED OF MORE THAN THE THREE-PLUS-ONE DIMENSIONS WE OBSERVE IN EVERYDAY CIRCUMSTANCES.

Higher-order dimensions are postulated to exist but are "curled up" and hidden away from us, much like the two-dimensional nature of a drinking straw is hidden when we look at it from far enough away (when it appears as a one-dimensional line). The story then goes that gravity is so weak because gravitons disappear into these curled-up dimensions, never to be seen (or, rather, felt) again.

In the high-energy collisions of the LHC, however, these dimensions would uncurl, preventing gravitons from disappearing and making gravity as strong as the other forces. The altered strength of gravity would then make it possible for **micro-black holes** to be created …

… NOT BECAUSE THERE WAS A LOT OF MASS PRESENT (LIKE WITH A GALACTIC BLACK HOLE), BUT BECAUSE THE LAWS OF PHYSICS HAD CHANGED IN THE EXTREME CONDITIONS OF THE LHC PROTON-PROTON COLLISIONS.

This confusion hilariously led to apocalyptic headlines in the run-up to the LHC switch-on, with people worrying about the earth being swallowed up in an artificial black hole. Theoretical physicist **Stephen Hawking** (b. 1942) wasn't worried about that, however:

THE BLACK HOLE WOULD DISAPPEAR IN A PUFF OF HAWKING RADIATION – AND I WOULD GET A NOBEL PRIZE.

While it was hoped that the LHC would smash down the door to a new world of physics beyond quarks, leptons, and the three forces that hold them together, no evidence for anything new has been found yet. What most people won't know, however, is that actually particle physicists have already unlocked physics beyond the Standard Model with a particle that is, in all fairness, very hard to notice.

The solar neutrino problem

It's easy to forget, with all of the attention particle colliders receive, that fundamental particle physics also encompasses some delightfully subtle experiments. It's the neutrino that has provided the first significant post-Standard Model surprise.

In the Standard Model, neutrinos are massless particles – they have to be, for reasons relating to parity and the way the weak force works. And yet predictions made using these massless Standard Model neutrinos were found to disagree with experimental observations …

... OBSERVATIONS NOT OF ACCELERATOR-MADE PARTICLES, BUT THOSE OF AN EXTRA-TERRESTRIAL ORIGIN: OUR OWN SUN. IT WAS THE *SOLAR NEUTRINO PROBLEM*.

The conundrum was this. The sun, as a constantly-burning nuclear furnace, was thought to pump out huge numbers of neutrinos – specifically, electron neutrinos (remember, there are three types in the Standard Model). In the late 1960s, **Raymond Davis Jr.** (1914–2006) and **John Bahcall** (1934–2005) attempted to measure these electron neutrinos.

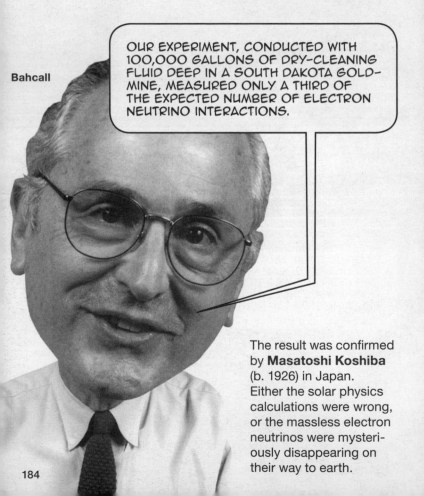

Bahcall

OUR EXPERIMENT, CONDUCTED WITH 100,000 GALLONS OF DRY-CLEANING FLUID DEEP IN A SOUTH DAKOTA GOLD-MINE, MEASURED ONLY A THIRD OF THE EXPECTED NUMBER OF ELECTRON NEUTRINO INTERACTIONS.

The result was confirmed by **Masatoshi Koshiba** (b. 1926) in Japan. Either the solar physics calculations were wrong, or the massless electron neutrinos were mysteriously disappearing on their way to earth.

184

It took just over 30 years to solve the problem. The first clue came from **Bruno Pontecorvo** (1913–93), who suggested that neutrinos might be changing into the other flavours – into muon neutrinos or tau neutrinos – as they traversed the solar system. The first hints of **neutrino oscillations** came from Japan in 1998, but definitive proof was obtained from the Canadian Sudbury Neutrino Observatory (SNO) in 2001.

WE USED HEAVY WATER IN OUR DETECTOR VOLUME TO SHOW THAT THE TOTAL NUMBER OF CALCULATED SOLAR NEUTRINOS WAS CORRECT, AND THAT TWO-THIRDS WERE INDEED MUON NEUTRINOS OR TAU NEUTRINOS.

THE SOLAR NEUTRINO PROBLEM WAS SOLVED.

While this was a great result for astrophysicists, the particle physicists were now stumped. In order to oscillate quantum-mechanically between flavours, as the quarks do, neutrinos needed to have a mass. This was not catered for by the Standard Model. Precision measurements are now under way at nuclear reactors and neutrino beam and cosmic neutrino experiments to further understand the nature of the oscillations.

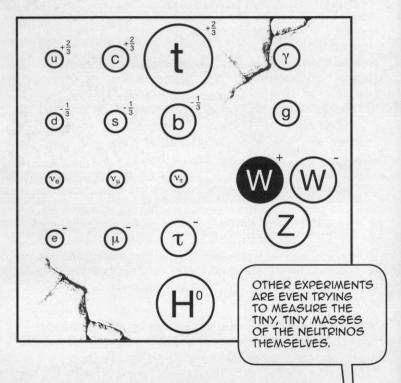

OTHER EXPERIMENTS ARE EVEN TRYING TO MEASURE THE TINY, TINY MASSES OF THE NEUTRINOS THEMSELVES.

Since their inception all those years ago, neutrinos have forced physicists to think about the very foundations of their theories – and this is still very much the case.

The journey continues

The type of experiment needed to study neutrinos – careful, precision measurements of ghost-like particles that simply don't want to be detected – is in sharp contrast to the brute-force collider approach represented by the LHC. But if physics beyond the Standard Model features new particles that have masses corresponding to energies we simply cannot achieve with our accelerators – such as supersymmetric particles, extra-dimensional resonances or superstrings – this may be our best strategy. We may have to look for new particles not through their creation, but through their subtle influence on well-known particle processes measured with extreme precision. Or we may have to search for supersymmetry in the form of stray particles from the galactic dark matter wind, following the neutrino physicists down mines with vast, silent detectors that can detect the delicate kiss of a sparticle on a single nucleus.

Perhaps the most exciting development has been the return to cosmic-ray physics. While the problems of experimental control remain, cosmic rays with energies thousands of times greater than anything the LHC can produce often bombard the earth. The trick is measuring them: cloud chambers and emulsions are no longer enough. The modern-day particle-hunter now has to lay their traps over vast tracts of land to snare their prey; for example, the Pierre Auger Observatory in Argentina has measured ultra-high-energy cosmic rays using an array of detectors spread over 3,000 square kilometres.

That said, it would be foolish to underestimate what we can achieve through engineering. The LHC will be upgraded to higher energies. At the risk of instantly dating this book, a linear collider capable of precision measurements at new energy frontiers will probably be built to complement the work of humanity's greatest machine. The real hope is that new technology – driven by the needs of fundamental, blue-sky research – will make energies at the tera-electron-volt scale not only accessible, but affordable. An LHC in every home would probably be a little ridiculous, but a desk-top Higgs boson factory would be something to aim for in the next few decades.

Such goals are grand, of course, but this is what we have seen throughout the particle physics journey: the ingenuity, the bravery, the tenacity, and the sheer stubbornness exercised by scientists – by people – in refusing to let matter triumph over mind. We may never find a complete Theory of Everything; indeed, philosophers may argue that such a feat is impossible. Let them think that. What I hope I have shown you throughout the course of this book is that particle physics isn't about the *what* of our theories – it's about the *how we got there* of them. It's not about the quarks or the leptons, the fermions or the bosons – it's the fact that we can dream up these ideas and put them to the test in dedicated laboratories that harness our understanding of matter and forces to further our understanding of matter and forces. It's a task that persistently tries our collective skills and intelligence – that continuously challenges our picture of the world – and yet it's a task that thousands of people from countries all around the world are still drawn to. It's a task that unites us through a shared curiosity as to the nature of reality. And that, perhaps, is the most important lesson we can learn from what we have achieved in particle physics so far.

By trying to find out what we're made of, we really have found out just what we're made of.

Acknowledgements

I am indebted to Duncan Heath and Oliver Pugh for (respectively) editing and illustrating this book; gentlemen, it was an utter joy and a pleasure. I must also thank CERN, STFC, James Gillies, Andrew Pontzen for their help and advice and, above all, Jen Whyntie for her love and support.

Dr T. Whyntie completed his PhD working on the CMS experiment at CERN, searching for supersymmetry. He didn't find it. He is currently a Visiting Academic at the Particle Physics Research Centre at Queen Mary, University of London.

Glossary

Alpha (particle): Type of radiation composed of helium nuclei; two protons, two neutrons (electric charge +2).

Anti-neutron: Antimatter equivalent of the neutron (no electric charge).

Anti-proton: Antimatter equivalent of the proton (electric charge −1).

Atom: Once thought to be the indivisible unit of matter, atoms actually consist of a nucleus containing Z protons and A − Z neutrons surrounded by Z electrons such that the atom has no overall charge (Z is the atomic number, A is the atomic mass).

Atomic mass: Mass of an atom's nucleus; in Atomic Mass Units (the mass of one proton), the sum of the number of protons and neutrons.

Atomic number: Number of protons in an atom's nucleus.

Baryon: Hadron composed of three quarks (or three anti-quarks).

Beta (particle): Type of radiation composed of an energetic electron or positron (electric charge −1 or +1 respectively).

Boson: Particle with an integer value of quantum mechanical spin, named for Satyendra Nath Bose (1894–1974).

Charge parity (CP): Symmetry in which the electric charge and the parity of a particle are transformed together.

Classical physics: Generally refers to physics pre-relativity and/or pre-quantum mechanics, i.e. 19th-century physics.

Collimation: The act of producing a narrow beam of radiation, usually achieved by placing blocking materials (the collimator) near the beam source.

Eightfold Way: System of organizing mesons and baryons by charge, strangeness and spin that predicted the existence of new particles and aided with the development of the quark model (see p. 108).

Electromagnetic force: One of the four fundamental forces.

Electron: Elementary particle in the first generation of the lepton family with a mass 1/1836 of the proton and a quantum-mechanical spin of one-half (electric charge −1).

Empiricism: School of thought that says knowledge of the world comes from experience and evidence.

Fermion: Particle with a half-integer value of quantum-mechanical spin, named for Enrico Fermi.

Gamma rays: Very high-energy electromagnetic radiation often observed as part of radioactive decay.

General Relativity: Theory in which mass and energy curve spacetime to produce what we observe as the gravitational force. See *Introducing Relativity* for more information.

Gluon: Exchange boson of the strong force.

Gravitational force: One of the four fundamental forces that describes the interaction of bodies that have (gravitational) mass. One of the major failings of the Standard Model is its inability to describe gravity.

Graviton: Hypothetical exchange boson of the gravitational force, thought to have spin 2.

Hadron: Elementary particle composed of quarks held together by the strong force.

Heisenberg's Uncertainty Principle: A feature in quantum mechanics that places a limit on the precision with which you can know the value of an observable quantity at the same time. For example, the more you know about a particle's position, the less you can know about its momentum, and vice versa.

Higgs boson: Spin-zero exchange boson associated with the Higgs field, responsible for imbuing certain particles with mass in the Standard Model (see pp. 144–9, 174–6).

Hyperon: Baryon containing a strange quark (not actually mentioned in the text, but included here for completeness).

Ion: An atom or molecule that has lost or gained electrons to give it an overall electric charge.

Ionization: The process by which atoms or molecules become ions (i.e. lose or gain electrons).

Isotope: For a given element, different isotopes have the same atomic number but different atomic mass; i.e. they will have the same number of protons but a different number of neutrons in the nucleus.

Kaon: A meson where one of the quarks is strange (or anti-strange).

Lepton: Class of spin-half elementary particles that experience all of the fundamental forces except the strong force (see p. 122).

Matrix mechanics: Formulation of quantum mechanics using matrices – arrays of numbers – to represent observable physical quantities.

Meson: Hadron consisting of a quark and an anti-quark (see p. 82).

Molecule: Electrically neutral particle made of one or more atoms held together by covalent (electron-sharing) bonds.

Muon: The second-generation charged lepton (see pp. 84–5).

Neutrino: Uncharged lepton that interacts only via the weak force (though recent results show they do have a very small mass). There are three flavours of neutrinos, corresponding to the charged leptons: the electron neutrino ν_e, the muon neutrino ν_μ and the tau neutrino ν_τ.

Neutron: Neutral, spin-half hadron consisting of one up quark and two down quarks.

Nucleons: Protons and neutrons (i.e. those that make up the nucleus).

Nucleus: The centre of the atom, containing its protons and neutrons.

Omega (baryon): Composed of three strange quarks, the Ω^- baryon was predicted by the Eightfold Way, which in turn helped with the development of the quark model.

Parity: A symmetry of nature relating to the inversion of an odd number of spatial coordinates (see p. 101).

Parton: Feynman's name for sub-nucleonic particles, i.e. quarks and gluons.

Photon: Spin-one exchange boson of the electromagnetic force.

Pion (a.k.a. pi meson): Hadron consisting of two up or down quarks; depending on the exact quark content, pions can be positively-charged, negatively-charged or neutral.

Positron: The antimatter equivalent of the electron (electric charge +1).

Proton: Positively-charged, spin-half hadron consisting of two up quarks and one down quark.

Quantization: The division of a quantity into discrete (i.e. non-continuous) values.

Quantum chromodynamics (QCD): Quantum field theory describing the interactions of quarks and gluons (see pp. 118–21).

Quantum electrodynamics (QED): Quantum field theory describing the interactions of charged matter and photons (see pp. 76–81).

Quantum Field Theory (QFT): A mathematical model in which every point in space is associated with an infinite number of tiny springs, each representing a potential chunk of energy. All matter and forces may be thought of as ripples passing through this network of springs.

Quark: Spin-half elementary particle that experiences all four fundamental forces (see pp. 107–17).

Radioactivity: The processes by which atomic nuclei undergo some sort of decay, losing energy via the release of ionizing radiation.

Renormalization (cf. QFT): Mathematical technique used to remove infinities from quantum field theories by absorbing them into measurable quantities (see p. 78).

Special Relativity: Theory that works from two postulates: that the laws of physics are the same wherever you test them, and that the speed of light is a fixed constant of the universe. See *Introducing Relativity* for more information.

Spin: Quantum-mechanical property of a particle relating to a sort of internal angular momentum that is almost entirely unlike spinning about its own axis. See *Introducing Quantum Theory* for a more thorough explanation!

Standard Model: The quantum field theory that combines our knowledge of the strong, weak and electromagnetic forces that describe the interactions of the known fundamental particles (see p. 104 onwards).

Strangeness: Before the quark model was formulated, this was the property assigned to particles containing strange quarks.

Strong force: One of the four fundamental forces of nature. It is described by quantum chromodynamics. Only particles with colour charge experience the strong force.

Supersymmetry: A hypothetical symmetry of nature that relates fermions and bosons.

Tau (lepton): The third-generation charged lepton (see p. 122). Not to be confused with the Tau of the Tau-Theta puzzle (see below).

Tau-Theta puzzle: The two decay modes of the positive kaon (also known as the K+) have differing final state parities. When both were observed in cosmic rays, it was thought that they must be due to the decay of two different particles, even though the mass and charge were the same. The puzzle was resolved when it was realized that parity could be violated, i.e. the parity of the final state did not have to be the same as the parity of the initial state (see pp. 100–03).

Wave mechanics: Formulation of quantum mechanics where a particle's properties (position, momentum, etc.) are described in terms of statistical distributions that resemble the mathematical description of waves.

Weak force: One of the four fundamental forces of nature. It is mediated by the massive W and Z bosons. Both quarks and leptons experience the weak force.

X-ray: High-energy electromagnetic radiation.

Index